Problem Solving Guide and Workbook

Problem Solving Flowcharts, Examples, and Practice Problems for Developing Problem Solving Skills

for

Introductory Chemistry

Second Edition

by

Steve Russo
Cornell University

Mike Silver
Hope College

Saundra Yancy McGuire
Louisiana State University

San Francisco • Boston • New York
Cape Town • Hong Kong • London • Madrid • Mexico City
Montreal • Munich • Paris • Singapore • Syndey • Tokyo • Toronto

To the many chemistry students for whom the development of problem-solving skills provides the greatest challenge and the greatest satisfaction during their study of introductory chemistry.

Executive Editor: Ben Roberts
Senior Developmental Editor: Margot Otway
Marketing Managers: Christy Lawrence
Director of Marketing: Stacey Treco
Associate Project Editor: Lisa Leung
Publishing Associate: Tony Asaro
Production Coordination: Joan Marsh
Production Management: Joan Keyes, Dovetail Publishing Services
Cover Image: Quade Paul, fiVth.com
Cover Design: Blakeley Kim
Manufacturing: Vivian McDougal
Printer and Binder: Victor Graphics

ISBN 0-321-06866-1

8 9 10—BRG—05
www.aw.com/bc

Table of Contents

To the Student

Most beginning chemistry students have difficulty solving problems because they are unfamiliar with the specific processes required to solve these problems. Successful problem solving in chemistry requires that you develop a systematic plan for solving the problem *and* that you use the appropriate mathematical procedures, when required, to arrive at the correct answer. (In this book the term *problem* refers to any question requiring a solution, whether arriving at the solution requires a mathematical procedure or not.)

This *Problem Solving Guide and Workbook* will help you develop your problem solving skills by explicitly demonstrating the procedures required for solving various types of problems presented in each chapter of the textbook. The flowcharts presented in the *Problem Solving Guide and Workbook* and in the textbook provide a specific set of steps that can be used to arrive at the correct solution for each type of problem presented. The *Problem Solving Guide and Workbook* provides additional examples and flowcharts for problems presented in the textbook. You should study these charts carefully, making sure you understand why and how each step is done. You should also know, however, that there are a variety of ways in which chemistry problems can be solved, and that each chart presents only one method. It is perfectly fine if you have a different procedure for solving a type of problem, provided your procedure is systematic and consistently gives the correct answer. You can even draw your own flowchart for the procedure that you have developed. If you develop your own procedure for solving certain problems, you should confer with your instructor to make sure that your procedure is correct.

The ability to solve a wide range of problems comes only with extensive practice. There is absolutely no substitute for working through as many problems as possible to sharpen your problem-solving skills. Therefore, this *Problem Solving Guide and Workbook* provides additional practice problems, practice quizzes, and cumulative quizzes after every three chapters to give you more experience with problem-solving tasks. Answers to selected practice problems and quizzes are provided in the back of the book so that you can tell whether you have correctly solved the problems. You should resist the temptation to look at the answer before you have earnestly attmpted to solve the problem because doing so will be detrimental to developing your problem-solving skills. A mathematics review is also provided in the Appendix for students who have not mastered all of the mathematical tools needed for problem solving in introductory chemistry. You should use the appendix to review the mathematical procedures needed to solve the problems in the textbook.

If you use the *Problem Solving Guide and Workbook* consistently, you should be successful in solving the problems in *Introductory Chemistry*. You may even find that you enjoy solving problems! I welcome comments and suggestions for improvement of this book, and I wish you success with your problem-solving tasks.

Saundra Yancy McGuire
Louisiana State University
smcgui1@lsu.edu

Acknowledgments

I extend my heartfelt gratitude to the many individuals who have worked tirelessly to produce this book.

My most sincere thanks go first to Melissa Bailey Crawford, who served as a writing assistant, problem solver, editorial consultant, and general motivator on this project. I could not have completed it without her. My thanks also to Steve Russo and Mike Silver, authors of the textbook, for their decision to include me in this project. I also wish to thank Ben Roberts, executive editor, for his continous encouragement and support; Maureen Kennedy, acquisitions editor, for her valuable assistance; and Lisa Leung, assistant editor, for her prompt and helpful responses to numerous queries during the writing process. Very special thanks go to Joan Keyes and Jonathan Peck of Dovetail Publishing Services for the meticulous manner in which they produced the finished work. Special thanks are also extended to the copy editor, Jan McDearmon; the accuracy reviewer, Yuri Zhorov; and the content reviewers, Paris Svoronos, Soraya Svoronos, Simon Bott, and Melissa Armstrong. Their corrections, comments, and suggestions improved the content and presentation immensely.

Last, but certainly not least, I want to thank my husband, Steve, for his patience, love, and support throughout this project.

CHAPTER

1

What Is Chemistry?

Overview: What You Should Be Able to Do

Chapter 1 provides an overview of matter and its transformations, and describes the steps in the scientific method. After mastering Chapter 1, you should be able to solve the following types of problems:

1. Identify a sample of matter as an elemental substance, a compound, a homogeneous mixture, or a heterogeneous mixture.
2. Indicate whether a given transformation of matter is a chemical or a physical transformation.

Chart 1.1 Identifying a sample of matter as an elemental substance, a compound, a homogeneous mixture, or a heterogeneous mixture

Step 1

- Determine whether the sample appears to consist of more than one substance. If more than one substance is visible, the sample is a heterogeneous mixture. If only one substance is visible, proceed to step 2.

⬇

Step 2

- Determine the composition of the sample (provided by the information in the problem). If only one element is present, the sample is an element; if only one compound is present, the sample is a compound. If the sample contains two or more substances, it is a homogeneous mixture.

Example 1

Classify a sample containing a tablespoon of salt dissolved in a cup of water as an elemental substance, a compound, a homogeneous mixture, or a heterogeneous mixture.

Solution

Perform step 1:

Only one substance is visible in a mixture containing a tablespoon of salt dissolved in a cup of water. We therefore proceed to step 2.

Perform step 2:

Because the composition of the sample is two substances (salt and water), the sample is a homogeneous mixture.

Example 2

Would a scoop of vanilla ice cream with chocolate chips be classified as an elemental substance, a compound, a homogeneous mixture, or a heterogeneous mixture?

Solution

Perform step 1:

Because it is visibly obvious that a scoop of vanilla ice cream with chocolate chips contains at least two substances, the sample is a heterogeneous mixture.

Example 3

Would a gold bar be classified as an elemental substance, a compound, a homogeneous mixture, or a heterogeneous mixture?

Solution

Perform step 1:

Only one substance is visible in a gold bar. We therefore proceed to step 2.

Perform step 2:

Because the composition of the sample is gold only (an element), the sample is an elemental substance.

Practice Problems

Indicate whether each of the following is an elemental substance, a compound, a homogeneous mixture, or a heterogeneous mixture.

1.1 a bowl of peaches and cream

1.2 a bar of zinc metal

1.3 a cup of distilled water

1.4 a tablespoon of maple syrup

Chart 1.2 **Indicating whether a given transformation of matter is a chemical or a physical transformation**

Step 1

- Determine whether the transformation results in a change in the physical state of the substance or in the production of a new substance.

⬇

Step 2

- Classify the transformation as physical if there is simply a change in the physical state, or classify the transformation as chemical if there is a new substance produced (evidence of a new substance may include an odor change, a color change, or fizzing).

Example 1

Classify the melting of a bar of platinum to form molten platinum as a physical transformation or a chemical transformation.

Solution

Perform step 1:

The transformation results in a change in the physical state of platinum.

Perform step 2:

Because the transformation results in only a change in physical state, a physical transformation has occurred.

Example 2

Classify the burning of a block of wood as a physical transformation or a chemical transformation.

Solution

Perform step 1:

The transformation results in the production of a new substance.

Perform step 2:

Because a new substance is produced, a chemical transformation has occurred.

Practice Problems

Indicate whether each of the following is a physical transformation or a chemical transformation.

1.5 the combination of H_2 and O_2 to form water

1.6 vaporizing a sample of liquid mercury

1.7 dissolving a tablespoon of sugar in water

1.8 the spoiling of a carton of orange juice

Quiz for Chapter 1 Problems

Note: The chapter quiz items involve problem solving and may require additional information provided in the textbook and the *Study Guide and Selected Solutions.*

1. Classify the following as elemental substances, compounds, homogeneous mixtures, or heterogeneous mixtures:

   (a) a glass of ice tea with lemon slices

   (b) a tablespoon of instant ice tea dissolved in water

   (c) a cup of homogenized milk

   (d) a beaker of liquid mercury

2. Indicate whether the change described below results in a physical transformation or a chemical transformation.

   (a) A block of iron metal is heated until it becomes a liquid.

   (b) An iron car rusts in an open field.

   (c) A sample of liquid bromine is converted to vapor.

3. Complete the following statement:

   In a chemical transformation, a ____________ ____________ is produced.

4. Identify the following as elemental substances or compounds:

   (a) sulfur dioxide (b) helium gas (c) ammonia (d) sulfur

5. Classify each of the following transformations as physical or chemical:

   (a) sublimation (b) burning (c) fermentation (d) condensation

6. Explain the difference between a heterogeneous mixture and a homogeneous mixture.

7. When a carton of milk spoils, a chemical transformation has occurred. Explain what indicates that this is a chemical transformation and not a physical transformation.

8. Give an example of each of the following:

   (a) an elemental substance (b) a compound
   (c) a homogeneous mixture (d) a heterogeneous mixture

9. Indicate whether the following statement is true or false. If false, explain why it is false.

   A glass of fizzing seltzer water is a homogeneous mixture.

10. Complete the following statement:

    When water is electrolyzed to produce hydrogen gas and oxygen gas, a _______________ transformation has occurred.

CHAPTER

2

The Numerical Side of Chemistry

Overview: What You Should Be Able to Do

Chapter 2 provides an introduction to the quantitative aspects of chemistry, including topics such as interpreting data, expressing numbers in different types of notation, converting quantities from one unit to another, and quantifying energy. After mastering Chapter 2, you should be able to solve the following types of problems:

1. Determine the number of significant figures in a numerical value.
2. Convert a number from scientific notation to normal notation.
3. Convert a number from normal notation to scientific notation.
4. Express the answer to the correct number of significant figures when multiplying or dividing measured values.
5. Express the answer to the correct number of significant figures when adding or subtracting measured values.
6. Express the answer to the correct number of significant figures when performing calculations involving both multiplication/division *and* addition/subtraction.
7. Use unit analysis to convert a measurement to another set of units.
8. Rearrange algebraic equations to solve for any variable that appears in the equation.

Chart 2.1 Determining the number of significant figures in a numerical value

Step 1

- Count the number of digits beginning with the first nonzero digit and ending with the last nonzero digit. (Zeros *before* the first nonzero digit are never significant, but all zeros between the first and last nonzero digit are significant.)

⬇

Step 2

- Determine whether the zeros after the last nonzero digit (if there are any) are significant by using the following rule:

> Trailing zeros after a decimal point are significant; trailing zeros before a decimal are significant only if the decimal point is written at the end of the number. (If the decimal point is not written at the end of the number, the trailing zeros are not significant.)

⬇

Step 3

- Determine the number of significant figures in the value by adding the significant digits from steps 1 and 2.

Example 1

What is the number of significant figures in 0.002 030?

Solution

Perform step 1:

There are three digits beginning with the 2 (the first nonzero digit) and ending with the 3 (the last nonzero digit).

Perform step 2:

The one trailing zero is significant because it occurs after the decimal point.

Perform step 3:

The number of significant figures in 0.002 030 is four (the sum of the three digits from step 1 and the one trailing zero from step 2).

Example 2

Determine the number of significant figures in 1.8000×10^5.

Solution

Perform step 1:

There are two digits beginning with the 1 (the first nonzero digit) and ending with the 8 (the last nonzero digit).

Perform step 2:

The three trailing zeros are significant because they occur after the decimal point.

Perform step 3:

The number of significant figures in 1.8000×10^5 is five (the sum of the three digits from step 1 and the three trailing zeros from step 2).

Example 3

The number 265,000,000 has how many significant figures?

Solution

Perform step 1:

There are three digits beginning with the 2 (the first nonzero digit) and ending with the 5 (the last nonzero digit).

Perform step 2:

The six trailing zeros are not significant because a decimal point is not written at the end of the number.

Perform step 3:

The number of significant figures in 265,000,000 is three (the sum of the three digits from step 1 and no significant trailing zeros from step 2).

Example 4

The number 265,000,000. has how many significant figures?

Solution

Perform step 1:

There are three digits beginning with the 2 (the first nonzero digit) and ending with the 5 (the last nonzero digit).

Perform step 2:

The six trailing zeros are significant because the decimal point has been written in at the end of the number.

Perform step 3:

The number of significant figures in 265,000,000. is nine (the sum of the three digits from step 1 and the six significant trailing zeros from step 2).

Example 5

How many significant figures are there in 856.612?

Solution

Perform step 1:

There are six digits beginning with the 8 (the first nonzero digit) and ending with the 2 (the last nonzero digit).

Perform step 2:

There are no trailing zeros.

Perform step 3:

The number of significant figures in 856.612 is six (the sum of the six digits from step 1 and no trailing zeros).

Practice Problems

Determine the number of significant figures in each numerical value below. (Assume all values are measurements.)

2.1 0.0035

2.2 8.302 150 0

2.3 1205.0

2.4 9365

2.5 8.150×10^{-3}

2.6 560,000.

2.7 0.005 600

2.8 500

2.9 0.500

Chart 2.2 Converting a number from scientific notation to normal notation

Step 1

- Determine whether the exponent on the power of 10 is positive or negative.

⬇

Step 2

- If the exponent is positive, move the decimal point to the right to produce the number in normal notation. If the exponent is negative, move the decimal point to the left to produce the number in normal notation. (Add zeros if there are not enough places to move the decimal point the appropriate number of places.)

Example 1

Convert 8.79×10^4 to normal notation.

Solution

Perform step 1:

The exponent on the power of 10 is positive (4).

Perform step 2:

Move the decimal point four places to the right to produce the number in normal notation. (Two zeros must be added to allow the decimal to be moved four places.) Therefore

$$8.79 \times 10^4 = 87{,}900$$

Example 2

Convert 4.3×10^{-2} to normal notation.

Solution

Perform step 1:

The exponent on the power of 10 is negative (-2).

Perform step 2:

Move the decimal point two places to the left to produce the number in normal notation. (One zero must be added to allow the decimal to be moved two places.) Therefore

$$4.3 \times 10^{-2} = .043, \text{ which is written as } 0.043$$

Practice Problems

Convert the following numbers, which are expressed in scientific notation, to numbers expressed in normal notation.

2.10 6.74×10^{-3}

2.11 6.74×10^{3}

2.12 $8.3\,421 \times 10^{9}$

2.13 3.58×10^{-5}

Chart 2.3 Converting a number from normal notation to scientific notation

Step 1

- Move the decimal point to the right or to the left until the decimal is located just after the first nonzero digit.

⬇

Step 2

- Determine the value of the exponent (x) on the power of ten. x reflects the number of places the decimal point was moved. If the decimal point was moved to the left, x is positive; if the decimal point was moved to the right, x is negative.

Example 1

Convert 0.002 95 to scientific notation.

Solution

Perform step 1:

The decimal must be moved three places to the right so that it will be located just after the first nonzero digit.

$$0.002\ 95 \text{ becomes } 2.95 \times 10^x$$

Perform step 2:

The x is equal to -3 because the decimal point was moved three places to the right. Therefore

$$0.002\ 95 = 2.95 \times 10^{-3}$$

Example 2

Convert 3,592,000. to scientific notation.

Solution

Perform step 1:

The decimal must be moved six places to the left so that it will be located just after the first nonzero digit. (Note that the three trailing zeros are significant because the decimal point is written at the end of the number.)

$$3{,}592{,}000. \text{ becomes } 3.592\ 000 \times 10^x$$

Perform step 2:

x is equal to 6 because the decimal point was moved six places to the left. Therefore

$$3{,}592{,}000. = 3.592\ 000 \times 10^6$$

Example 3

Convert 0.000 000 052 7 to scientific notation.

Solution

Perform step 1:

The decimal must be moved eight places to the right so that it will be located just after the first nonzero digit.

$$0.000\ 000\ 052\ 7 \text{ becomes } 5.27 \times 10^{x}$$

Perform step 2:

x is equal to -8 because the decimal point was moved eight places to the right. Therefore

$$0.000\ 000\ 052\ 7 = 5.27 \times 10^{-8}$$

Practice Problems

Convert the following numbers, which are expressed in normal notation, to numbers expressed in scientific notation.

2.14 83,000,000

2.15 0.000 782 00

2.16 503

2.17 0.467

2.18 29

2.19 .0006

Chart 2.4 Expressing the answer to the correct number of significant figures when multiplying or dividing measured values

Step 1

- Determine the number of significant figures in each number being multiplied or divided. The smallest number of significant figures in these values is the number of significant figures allowed in the answer.

⬇

Step 2

- Perform the arithmetic operation(s).

⬇

Step 3

- Round the answer to the correct number of significant figures. (The rules for rounding numbers are given in the Appendix at the back of this book.)

Example 1

Perform the following calculation, expressing the answer to the correct number of significant figures:

$$0.036 \times 2.30$$

Solution

Perform step 1:

There are two significant figures in 0.036, and there are three significant figures in 2.30. Therefore the answer can have only two significant figures.

Perform step 2:

$0.036 \times 2.30 = 0.0828$

Perform step 3:

There must be two significant figures in the answer, so 0.0828 must be rounded to 0.083. The answer can also be expressed as 8.3×10^{-2}.

Example 2

Perform the following calculation, expressing the answer to the correct number of significant figures:

$$\frac{5.876\ 50 \times 10^5}{0.002\ 41 \times 10^8}$$

Solution

Perform step 1:

There are six significant figures in $5.876\ 50 \times 10^5$, and there are three significant figures in $0.002\ 41 \times 10^8$. Therefore the answer can have only three significant figures.

Perform step 2:

$$\frac{5.876\ 50 \times 10^5}{0.002\ 41 \times 10^8} = 2.4384$$

Perform step 3:

There must be three significant figures in the answer, so 2.4384 must be rounded to 2.44.

Example 3

Perform the following calculation, expressing the answer to the correct number of significant figures:

$$\frac{0.9842 \times 0.389}{0.0812 \times 298.15}$$

Solution

Perform step 1:

There are four significant figures in 0.9842, three significant figures in 0.389, three significant figures in 0.0812, and five significant figures in 298.15. Therefore the answer can have only three significant figures.

Perform step 2:

$$\frac{0.9842 \times 0.389}{0.0812 \times 298.15} = 0.015\ 814$$

Perform step 3:

There must be three significant figures in the answer, so 0.015 814 must be rounded to 0.0158. The answer can also be expressed as 1.58×10^{-2}.

Practice Problems

Perform the following calculations and express the answers to the correct number of significant figures.

2.20 $\frac{2.35}{40.08}$ **2.21** $0.065 \times 6.023 \times 10^{23}$ **2.22** $\frac{2.500 \times 453.6}{1.21}$

2.23 $\frac{5.000 \times 0.001\ 300}{0.1256}$ **2.24** $\frac{2.0 \times 20.0}{0.080 \times 8.0}$

Chart 2.5 Expressing the answer to the correct number of significant figures when adding or subtracting measured values

Step 1

- Determine the number of decimal places (digits behind the decimal point) in each number being added or subtracted. The smallest number of decimal places in these values is the number of decimal places allowed in the answer.

↓

Step 2

- Perform the arithmetic operation(s).

↓

Step 3

- Round the answer to the correct number of decimal places.

Example 1

Perform the following calculation, expressing the answer to the correct number of significant figures:

$$7.32 + 0.0056$$

Solution

Perform step 1:

There are two decimal places in 7.32, and there are four decimal places in 0.0056. Therefore the answer can have only two decimal places.

Perform step 2:

$7.32 + 0.0056 = 7.3256$

Perform step 3:

There must be two decimal places in the answer, so 7.3256 must be rounded to 7.33.

Example 2

Perform the following calculation, expressing the answer to the correct number of significant figures: $92.75 - 16$

Solution

Perform step 1:

There are two decimal places in 92.75, and no decimal places in 16. Therefore the answer can have no decimal places.

Perform step 2:

$92.75 - 16 = 76.75$

Perform step 3:

There must be no decimal places in the answer, so 76.75 must be rounded to 77.

Example 3

Perform the following calculation, expressing the answer to the correct number of significant figures:

$$23.113 + 76.2 - 5.31$$

Solution

Perform step 1:

There are three decimal places in 23.113, one decimal place in 76.2, and two decimal places in 5.31. Therefore the answer can have only one decimal place.

Perform step 2:

$23.113 + 76.2 - 5.31 = 94.003$

Perform step 3:

There must be one decimal place in the answer, so 94.003 must be rounded to 94.0.

Practice Problems

Perform the following calculations and express the answers to the correct number of significant figures.

2.25 $1.351 + 12.0236$

2.26 $0.653 + 1.2$

2.27 $35 - 1.356$

2.28 $36.3 + 0.159 - 0.12$

2.29 $359 - 2.21 - 0.167$

Chart 2.6 Expressing the answer to the correct number of significant figures when performing calculations involving both multiplication/division *and* addition/subtraction

Step 1

- Perform the addition/subtraction step(s) and note the correct number of significant figures that the answer will have (after rounding to the correct number of decimal places), but do not round.

⬇

Step 2

- Perform the multiplication/division step(s) and note the correct number of significant figures allowed in the answer.

⬇

Step 3

- Round the answer to the correct number of significant figures.

Example 1

Perform the following calculation, expressing the answer to the correct number of significant figures:

$$\frac{18.6 + 0.071}{12.02}$$

Solution

Perform step 1:

$18.6 + 0.071 = 18.671$, which will become 18.7 when rounded to one decimal place. 18.7 has three significant figures.

Perform step 2:

We use the number before rounding from step 1 to do the calculation.

$$\frac{18.671}{12.02} = 1.553\ 327\ 8$$

There must be three significant figures in the answer because 18.7 has three significant figures and 12.02 has four significant figures.

Perform step 3:

1.553 327 8 must be rounded to three significant figures. The correct answer is 1.55.

Example 2

Perform the following calculation, expressing the answer to the correct number of significant figures:

$$(1.800)(99.6 - 32)$$

Solution

Perform step 1:

$99.6 - 32 = 67.6$, which will become 68 when rounded to no decimal places. 68 has two significant figures.

Perform step 2:

$(1.800)(67.6) = 121.68$

There must be two significant figures in the answer because 68 (from step 1) has two significant figures and 1.800 has four significant figures.

Perform step 3:

121.68 must be rounded to two significant figures. The correct answer is 120, or 1.2×10^1.

Example 3

Perform the following calculation, expressing the answer to the correct number of significant figures:

$$\frac{(0.7577 \times 34.969) + (0.2423 \times 36.966)}{6.0 \times 10^{23}}$$

Solution

Perform step 1:

In this case, we must first multiply the number in parentheses to determine which two numbers will be added. After doing the multiplication steps (and rounding to the correct number of significant figures), we obtain

$$\frac{26.50 + 8.957}{6.0 \times 10^{23}}$$

Now we add the two numbers. There are two decimal places in 26.50, and three decimal places in 8.957. Therefore the sum of these two numbers can have only two decimal places, but we will not round yet. We will use 35.457 in the next step.

Perform step 2:

$$\frac{35.457}{6.0 \times 10^{23}} = 5.8899 \times 10^{-23}$$

There must be two significant figures in the answer because 35.46 from step 1 (after rounding) has four significant figures, and 6.0×10^{23} has two significant figures.

Perform step 3:

5.8899×10^{-23} must be rounded to 5.9×10^{-23}.

Practice Problems

Perform the following calculations and express the answers to the correct number of significant figures.

2.30 $\dfrac{1.89 + 13.957}{5.1}$

2.31 $\dfrac{(0.632)(2.33 - 0.867)}{12}$

2.32 $\dfrac{(0.667)(109.57) + (0.333)(83.65)}{3.22}$

Chart 2.7 Using unit analysis to convert a measurement to another set of units

Step 1

- Write the measurement with its units.
- Identify the units to which you need to convert.

⬇

Step 2

- Multiply the measurement by a conversion factor that will let you cancel the units you don't want. (Conversion factors are given in the textbook.)
- Cancel units. You can cancel identical units on the top and bottom of the expression.
- If one conversion factor won't yield the units you need, multiply by additional conversion factors.

⬇

Step 3

- Perform the calculation.

Example 1

Convert 253 miles to centimeters.

Solution

Perform step 1:

253 miles must be converted to centimeters. However, the conversion factor we have available in the textbook will allow us to convert miles to kilometers.

Perform step 2:

We convert miles to kilometers, using the conversion factor 1 km = 0.621 37 mi. Therefore

$$253\ \cancel{\text{mi}} \times \frac{1\ \text{km}}{0.621\ 37\ \cancel{\text{mi}}} = 407\ \text{km}$$

Note that we could have used the conversion factor 1 mi = 1.6093 km in the following way:

$$253\ \cancel{\text{mi}} \times \frac{1.6093\ \text{km}}{1\ \cancel{\text{mi}}} = 407\ \text{km}$$

Notice that both conversion factors yield identical results.

We must now multiply by the additional conversion factors to convert kilometers to centimeters.

Perform step 3:

$$407\ \cancel{\text{km}} \times \frac{1000\ \cancel{\text{m}}}{1\ \cancel{\text{km}}} \times \frac{100\ \text{cm}}{1\ \cancel{\text{m}}} = 40{,}700{,}000\ \text{cm} = 4.07 \times 10^{7}\ \text{cm}$$

Example 2

How many grams of silver are there in 750.0 mL? The density of silver is 10.5 g/mL.

Solution

Perform step 1:

750.0 mL Ag

This must be converted to grams, using the density as the conversion factor.

Perform step 2:

$$750.0 \text{ mL Ag} \times \frac{10.5 \text{ g Ag}}{1 \text{ mL Ag}}$$

Perform step 3:

$$750.0 \text{ mL Ag} \times \frac{10.5 \text{ g Ag}}{1 \text{ mL Ag}} = 7875 \text{ g Ag}$$

7875 g Ag = 7880, or 7.88×10^3 g Ag to three significant figures

Example 3

How many liters of oxygen gas are occupied by 170.2 grams of oxygen at standard temperature and pressure? At standard temperature and pressure one gram of oxygen occupies 0.700 liters.

Solution

Perform step 1:

170.2 g oxygen

This must be converted to liters, using the conversion factor provided in the problem.

Perform step 2:

$$170.2 \text{ g oxygen} \times \frac{0.700 \text{ L oxygen}}{1 \text{ g oxygen}}$$

Perform step 3:

$$170.2 \text{ g oxygen} \times \frac{0.700 \text{ L oxygen}}{1 \text{ g oxygen}} = 119 \text{ L oxygen}$$

Example 4

Convert 70.0 miles per hour to kilometers per minute.

Solution

Perform step 1:

70.0 mi/h

This must be converted to km/min, using the appropriate conversion factors.

Perform step 2:

$$\frac{70.0\ \text{mi}}{1\ \text{h}} \times \frac{1.6093\ \text{km}}{1\ \text{mi}} \times \frac{1\ \text{h}}{60\ \text{min}}$$

Perform step 3:

$$\frac{70.0\ \text{mi}}{1\ \text{h}} \times \frac{1.6093\ \text{km}}{1\ \text{mi}} \times \frac{1\ \text{h}}{60\ \text{min}} = 1.88\ \text{km/min}$$

Example 5

Convert 15 ft^2 to cm^2.

Solution

Perform step 1:

15 ft^2 must be converted to cm^2. However, the conversion factor we have available will allow us to convert feet to inches, and inches to centimeters. Because the problem involves square feet and square centimeters, we will have to square each of our conversion factors.

Perform step 2:

We can convert ft^2 to cm^2, using the conversion factors 1 ft = 12 in. and 1 in. = 2.54 cm. Therefore

$$15\ \text{ft}^2 \times \left(\frac{12\ \text{in.}}{1\ \text{ft}}\right)^2 \times \left(\frac{2.54\ \text{cm}}{1\ \text{in.}}\right)^2 = 15\ \text{ft}^2 \times \left(\frac{12^2\ \text{in.}^2}{1^2\ \text{ft}^2}\right) \times \left(\frac{2.54^2\ \text{cm}^2}{1^2\ \text{in.}^2}\right) = 1.4 \times 10^4\ \text{cm}^2$$

Practice Problems

Perform the following conversions, using unit analysis, and express the answers to the correct number of significant figures.

2.33 100.0 grams of iron to liters of iron. The density of iron is 7.87 g/mL.

2.34 120 km/h to miles per minute

2.35 536 milligrams to kilograms

2.36 25 m^2 to cm^2

2.37 $25/$ft^2$ to dollars per square inch

Chart 2.8 Rearranging algebraic equations to solve for any variable that appears in the equation

Step 1

- Write the equation.
- Identify the variable that must be solved for in the problem.

⬇

Step 2

- Isolate this variable on one side of the equation by algebraically solving the equation for this variable. (Consult the mathematics review in the appendix if you are not sure how to do this.)
- If the variable is already isolated on one side of the equation, proceed to step 3.

⬇

Step 3

- Substitute the values given in the problem for the variables other than the one for which you are solving.
- Make sure that the values being substituted are in the correct units. If not, perform the appropriate conversions before substituting.

⬇

Step 4

- Perform the arithmetic operation(s) to arrive at the answer.

Example 1

Calculate the amount of heat (in joules) needed to raise the temperature of 550.0 g of iron by 25.0°C. The specific heat of iron is 0.449 J/g · °C. The algebraic equation relating the variables is shown below.

$$\text{Heat} = \text{Specific heat} \times \text{Mass of iron} \times \text{Change in temperature}$$

Solution

Perform step 1:

Heat = Specific heat × Mass of iron × Change in temperature

The variable to be solved for is heat, which is already isolated on the left side of the equation.

Perform step 3:

Heat = Specific heat × Mass of iron × Change in temperature

Heat = 0.449 J/g · °C × 550.0 g × 25.0°C

Perform step 4:

Heat = $0.449\ J/g \cdot °C \times 550.0\ g \times 25.0°C = 6.17 \times 10^3\ J$

Example 2

Calculate the change in temperature resulting from adding 3.134×10^5 J of heat to a 5.00 g block of aluminum. The specific heat of aluminum is 0.901 J/g · °C. The algebraic equation relating the variables is shown below.

$$\text{Heat} = \text{Specific heat} \times \text{Mass of aluminum} \times \text{Change in temperature}$$

Solution

Perform step 1:

Heat = Specific heat × Mass of aluminum × Change in temperature

The variable to be solved for is "Change in temperature," which must be isolated on one side of the equation.

Perform step 2:

"Change in temperature" can be isolated by dividing both sides of the equation by "Specific heat" and by "Mass of aluminum." The resulting equation is

$$\text{Change in temperature} = \frac{\text{Heat}}{\text{Specific heat} \times \text{Mass of aluminum}}$$

Perform step 3:

Note that before substituting the values, we must convert the mass from kilograms to grams, so that the mass unit is consistent with the units of specific heat.

$$\text{Change in temperature} = \frac{\text{Heat}}{\text{Specific heat} \times \text{Mass of aluminum}}$$

$$\text{Change in temperature} = \frac{3.134 \times 10^5\ J}{0.901\ J/g \cdot °C \times 5.00 \times 10^3 g}$$

Perform step 4:

$$\text{Change in temperature} = \frac{3.134 \times 10^5\ J}{0.901\ J/g \cdot °C \times 5.00 \times 10^3 g}$$

$$\text{Change in temperature} = 69.6°C$$

Practice Problems

2.38 How many joules are required to raise the temperature of 972.0 g of aluminum from 0.0°C to 50.0°C? The specific heat of aluminum is 0.901 J/g · °C.

2.39 25.0 kJ of heat is added to a 500.0-g bar of iron metal at 25.0°C. What is the final temperature of the iron bar? The specific heat of iron is 0.449 J/g · °C.

Quiz for Chapter 2 Problems

1. Indicate the number of significant figures in each measured number below.

   (a) 357 mL (b) 1.0600 L (c) 0.000 501 (d) 23,000 (e) 0.087 201 00

2. Indicate the number of zeros that are significant in each measured number below.

   (a) 0.003 810 0 (b) 500.00 (c) 0.030 010 (d) 5200. (e) 5200

3. Convert the following numbers from scientific notation to normal notation:

   (a) 8.59×10^{-3} (b) 2.76×10^{2} (c) 2.76×10^{-2} (d) 7.2×10^{9}

4. Convert the following numbers from normal notation to scientific notation:

   (a) 0.000 000 000 830 4 (b) 9,500,000 (c) 0.013 (d) 58.3 (e) 0.583

5. Perform the following arithmetic operations and express the answer to the correct number of significant figures:

   (a) $0.392 + 51.4$ (b) $0.001 + 5.32$ (c) $273.15 - 28.3$ (d) $1582 + 0.59$

6. Perform the following arithmetic operations and express the answer to the correct number of significant figures:

   (a) 8.63×0.58 (b) $\frac{6.02}{3.0}$ (c) $\frac{5260 \times 12.0}{2.1}$ (d) $\frac{22.4 \times 3.1}{0.3214}$

7. Perform the following arithmetic operations and express the answer to the correct number of significant figures:

   (a) $(8.39 \times 4.7) + 6.23$ (b) $\frac{6.8 \times 10^{5}}{7.31 - 4.2}$ (c) $\frac{3.90 \times 8.631}{6.0}$ (d) $\frac{3.9 \times 8.631}{6}$

8. Use unit analysis to perform each of the following conversions:

   (a) 58.3 mL to L

   (b) 0.004 63 cm to m

   (c) 839 kg to g

   (d) 92.0 $in.^2$ to cm^2

9. Copper has a density of 8.96 g/mL. Calculate the volume occupied by 125.0 g of copper.

10. Mercury has a density of 13.6 g/mL. What is the weight of 75.3 mL of mercury?

11. Calculate the number of cubic inches in 5000.0 cm^3.

12. A carpet is advertised at a price of $30.00 per square foot. What is the cost of the carpet per square inch?

13. Calculate the heat (in joules) required to raise the temperature of 1000.0 grams of aluminum from 25.0°C to 50.0°C. (The specific heat of aluminum is 0.901 J/g · °C.)

14. A sample of iron experiences a temperature increase of 35.0°C when 5892 joules of heat are added to it. What is the mass of the iron? (The specific heat of iron is 0.449 J/g · °C.)

15. The maximum speed of some U.S. interstate highways is 75 mi/h. What is this speed in km/h?

16. How are the measured values 156,000. and 156,000 different? Which one assumes a more accurate measuring device?
17. The formula for converting temperature from °C to °F is given below.

$$°F = 1.80°C + 32$$

Convert 37°C to °F and express the answer to the correct number of significant figures.

18. Fill in the blanks with the words *greater than*, *less than*, or *equal to*.
   (a) 5.32 is _______ _______ 53.2×10^1.
   (b) 5.32 is _______ _______ 53.2×10^{-1}.
   (c) 5.32 is _______ _______ 53.2×10^0.
   (d) 5.32×10^{-1} is _______ _______ 0.532.
   (e) 5.32×10^{-1} is _______ _______ 53.2.
19. Calculate the density of a piece of metal that weighs 386 grams and occupies a volume of 20.0 milliliters.
20. The formula for converting °C to kelvins is shown below.

$$K = °C + 273.15$$

Convert 83.1°C to K, and express the answer to the correct number of significant figures.

CHAPTER

3

The Evolution of Atomic Theory

Overview: What You Should Be Able to Do

Chapter 3 provides an introduction to the structure of the atom, the periodic table, and the properties of the elements. After mastering Chapter 3, you should be able to solve the following types of problems:

1. Calculate the percent by mass of the elements in a compound when given the grams of each element in the compound.
2. Calculate the weighted average atomic mass of any element when given the percent abundance and the mass of each isotope of the element.
3. Arrange a group of elements in order of increasing (or decreasing) atomic radius (atomic size) or ionization energy.

Chart 3.1 Calculating the percent by mass of the elements in a compound when given the grams of each element in the compound

Step 1

- Determine the mass of each element in the compound from the information given.
- If all but one of the masses are given, the missing mass value can be obtained by subtracting the sum of the masses of the other elements from the total mass of the compound.

⬇

Step 2

- Determine the total mass of the compound.
- Add the masses of the elements in the compound if the total mass is not given in the problem.

⬇

Step 3

- Divide the mass of each element by the mass of the total compound, and multiply this number by 100% to obtain the percent abundance of each element.
- The sum of the mass percents for all elements should be very close to 100%. (Rounding may make the sum a little less or a little more than 100%.)

⬇

Step 4

- Add all of the mass percents to verify that they total a number very close to 100%.
- If the total is not close to 100% redo the problem, beginning at step 1, to correct your error.

Example 1

A 25.00-g sample of PCl_3 contains 5.64 g of phosphorus. Calculate the mass percent of P and Cl in PCl_3.

Solution

Perform step 1:

Mass of phosphorus = 5.64 g P

Mass of Cl = 25.00 − 5.64 = 19.36 g Cl

Perform step 2:

The total mass of the compound is 25.00 g.

Perform step 3:

$$\%\ P = \frac{5.64\ g\ P}{25.00\ g\ cmpd} \times 100\% = 22.6\%\ P$$

$$\%\ Cl = \frac{19.46\ g\ Cl}{25.00\ g\ cmpd} \times 100\% = 77.84\%\ Cl$$

Perform step 4:

Sum of mass percents: 22.6% P + 77.84% Cl = 100.4% (Close to 100%)

Example 2

Calculate the mass percent of each element in a compound containing aluminum and oxygen if 203.92 g of compound contains 105.8 g of Al.

Solution

Perform step 1:

Mass of aluminum = 105.8 g Al

Mass of oxygen = 203.92 − 105.8 = 98.1 g O

Perform step 2:

The total mass of the compound is 203.92 g.

Perform step 3:

$$\%\ \text{Al} = \frac{105.8\ \text{g Al}}{203.92\ \text{g cmpd}} \times 100\% = 51.88\%\ \text{Al}$$

$$\%\ \text{O} = \frac{98.1\ \text{g O}}{203.92\ \text{g cmpd}} \times 100\% = 48.1\%\ \text{O}$$

Perform step 4:

Sum of mass percents: 51.9% Al + 48.1% O = 100.0%

Example 3

Calculate the mass percent of each element in a compound containing 15.00 g sulfur and 22.44 g oxygen.

Solution

Perform step 1:

Mass of sulfur = 15.00 g S

Mass of oxygen = 22.44 g O

Perform step 2:

The total mass of the compound is 15.00 g S + 22.44 g O = 37.44 g cmpd.

Perform step 3:

$$\%\ \text{S} = \frac{15.00\ \text{g S}}{37.44\ \text{g cmpd}} \times 100\% = 40.06\%\ \text{S}$$

$$\%\ \text{O} = \frac{22.44\ \text{g O}}{37.44\ \text{g cmpd}} \times 100\% = 59.94\%\ \text{O}$$

Perform step 4:

Sum of mass percents: 40.06% S + 59.94% O = 100.00%

Practice Problems

3.1 Calculate the mass percent of each element in a compound containing carbon and sulfur if the compound contains 24.00 g of carbon and 128.00 g of sulfur.

3.2 Calculate the mass percent of each element in a compound containing sodium and oxygen if 39.00 g of compound contains 16.00 g of oxygen.

3.3 Calculate the mass percent of each element in a compound that contains 5.40 g of aluminum and 4.80 g of oxygen.

Chart 3.2 Calculating the weighted average atomic mass of any element when given the percent abundance and the mass of each isotope of the element

Step 1

- Determine the mass of each isotope in the compound from the information given.

⬇

Step 2

- Determine the percent abundance of each isotope in the compound from the information given.
- If all but one of the percent abundances are given, the missing percent abundance value can be obtained by subtracting the sum of the percent abundances of the other isotopes from 100%.

⬇

Step 3

- For each isotope divide the percent abundance by 100%, and multiply the result times the mass of the isotope.
- This result represents the portion of the weight of the element contributed by each isotope.

⬇

Step 4

- Add all of the portions obtained in step 3.
- This sum is the weighted average atomic mass of the element.

Example 1

Calculate the weighted average atomic mass for bromine (Br). Br consists of two isotopes, ^{79}Br (atomic mass 78.9183 amu, abundance 50.69%) and ^{81}Br (atomic mass 80.9163, abundance 49.31%).

Solution

Perform step 1:

Isotopic masses:

$^{79}Br = 78.9183$ amu

$^{81}Br = 80.9163$ amu

Perform step 2:

Percent abundance of each isotope:

$^{79}Br = 50.69\%$

$^{81}Br = 49.31\%$

Perform step 3:

Portion of the mass contributed by each isotope:

$$^{79}Br = \frac{50.69\%}{100\%} \times 78.9183 \text{ amu} = 40.00 \text{ amu from } ^{79}Br$$

$$^{81}Br = \frac{49.31\%}{100\%} \times 80.9163 \text{ amu} = 39.90 \text{ amu from } ^{81}Br$$

Perform step 4:

Add all portions from step 3:

Weighted average atomic mass for Br = 40.00 amu + 39.90 amu = 79.90 amu

Example 2

Calculate the weighted average atomic mass for magnesium (Mg). Mg consists of three isotopes, ^{24}Mg (atomic mass 23.9850 amu, abundance 78.99%), ^{25}Mg (atomic mass 24.9858 amu, abundance 10.00%), and ^{26}Mg (atomic mass 25.9826 amu).

Solution

Perform step 1:

Isotopic masses:

$^{24}Mg = 23.9850$ amu

$^{25}Mg = 24.9858$ amu

$^{26}Mg = 25.9826$ amu

Perform step 2:

Percent abundance of each isotope:

$^{24}Mg = 78.99\%$

$^{25}Mg = 10.00\%$

$^{26}Mg = 100\% - (78.99\% + 10.00\%) = 11.01\%$

Perform step 3:

Portion of the mass contributed by each isotope:

$$^{24}Mg = \frac{78.99\%}{100\%} \times 23.9850 \text{ amu} = 18.95 \text{ amu from } ^{24}Mg$$

$$^{25}Mg = \frac{10.00\%}{100\%} \times 24.9858 \text{ amu} = 2.499 \text{ amu from } ^{25}Mg$$

$$^{26}Mg = \frac{11.01\%}{100\%} \times 25.9826 \text{ amu} = 2.861 \text{ amu from } ^{26}Mg$$

Perform step 4:

Add all portions from step 3:

$$\text{Weighted average atomic mass for Mg} = 18.95 \text{ amu} + 2.499 \text{ amu} + 2.861 \text{ amu} = 24.31 \text{ amu}$$

Practice Problems

3.4 Boron (B) consists of two naturally occurring isotopes, ^{10}B and ^{11}B. If the abundance of ^{10}B is 19.9%, what is the abundance of ^{11}B?

3.5 Argon (Ar) consists of three naturally occurring isotopes. ^{36}Ar (atomic mass 35.9675 amu, abundance 0.3365%), ^{38}Ar (atomic mass 37.9627 amu, abundance 0.0632%), and ^{40}Ar (atomic mass 39.9634 amu, abundance 99.6003%). Calculate the weighted average atomic mass for Ar.

Chart 3.3 Arranging a group of elements in order of increasing (or decreasing) atomic radius (atomic size) or ionization energy

Step 1

- Determine the periodic trend for the specified property (e.g., atomic radius increases from top to bottom and decreases from left to right on the periodic table).

⬇

Step 2

- Note whether the problem specifies an arrangement in increasing order or decreasing order.
- If increasing order is specified, the smallest will be first; if decreasing order is specified, the largest will be first.

⬇

Step 3

- Arrange the elements in the appropriate order.

Example 1

Arrange the following elements in order of increasing atomic radius:

Mg, S, Si, P, Na

Solution

Perform step 1:

The atomic radius increases from left to right.

Perform step 2:

The problem asks for arranging in order of increasing atomic radius; therefore the smallest atom (the one farthest to the right in the periodic table) will be first. This is S.

Perform step 3:

$S < P < Si < Mg < Na$

Example 2

Arrange the following elements in order of decreasing ionization energy:

Se, O, Te, Po, S

Solution

Perform step 1:

The ionization energy decreases from top to bottom.

Perform step 2:

The problem asks for arranging in order of decreasing ionization energy; therefore the element with the largest ionization energy (the one closest to the top in the periodic table) will be first. This is O.

Perform step 3:

$O > S > Se > Te > Po$

Example 3

Arrange the following elements in order of decreasing atomic radius:

Sn, Rb, Xe, Sr, Te

Solution

Perform step 1:

The atomic radius decreases from left to right.

Perform step 2:

The problem asks for arranging in order of decreasing atomic radius; therefore the element with the largest atomic radius (the one closest to the left in the periodic table) will be first. This is Rb.

Perform step 3:

Rb > Sr > Sn > Te > Xe

Practice Problems

3.6 Arrange the following elements in order of increasing ionization energy:

Br, F, Cl, I

3.7 Arrange the following elements in order of decreasing atomic radius:

Br, F, Cl, I

3.8 Which element below has the smallest ionization energy?

Ge, Sn, Si, C

Quiz for Chapter 3 Problems

1. Carbonic acid has the formula H_2CO_3. Calculate the mass percents of the elements in carbonic acid.
2. Calculate the weighted average atomic mass for silver (Ag). Ag consists of two isotopes, ^{107}Ag (atomic mass 106.9051 amu, abundance 51.840%) and ^{109}Ag (atomic mass 108.9048, abundance 48.162%).
3. Arrange the following elements in order of increasing atomic radius:

   Pb, Po, Tl, Cs
4. Potassium consists of three naturally occurring isotopes, ^{39}K (atomic mass 38.9637 amu), ^{40}K (atomic mass 39.9640 amu), and ^{41}K (atomic mass 40.9618). If the weighted average atomic mass of potassium is 39.098, which of the three isotopes is present in the greatest amount in naturally occurring potassium? Explain your choice.

5. Arrange the following elements in order of increasing ionization energy:

   O, C, Ne, F, Be

6. Circle the element that has the greatest atomic radius.

   Al, In, Tl, B, Ga

7. Indium (In) consists of two naturally occurring isotopes, ^{113}In (atomic mass 112.9041 amu) and ^{115}In (atomic mass 114.9039 amu). The percent abundance of ^{113}In is 4.29%. What is the percent abundance of ^{115}In?

8. Arrange the following elements in order of decreasing atomic radius:

   Na, Rb, K, Cs, Li

9. Circle the element with the smallest ionization energy.

   Ar, Ne, He, Xe, Kr

10. Arrange the following elements in order of decreasing ionization energy:

    O, C, Ne, F, Be

Cumulative Quiz for Chapters 1, 2, and 3

1. Classify the following as elemental substances, compounds, homogeneous mixtures, or heterogeneous mixtures:
   (a) A chocolate chip cookie
   (b) A tablespoon of instant coffee dissolved in water
   (c) A tablespoon of sugar
   (d) A bar of iron
   (e) A cup of sweetened grape juice
2. Indicate whether the change described below results in a physical transformation or a chemical transformation.
   (a) A block of iron metal rusts in an open field.
   (b) An ice cube melts on a hot summer day.
   (c) Oxygen is converted to ozone in the upper atmosphere.
3. Indicate whether the following statement is true or false. If false, explain why it is false.

   S_8 is a compound because there is more than one element in the formula.
4. Complete the following statement:

   A(n) ___________ is a substance that is made up of only one type of atom, whereas a(n) ___________ is made up of atoms of two or more different substances.
5. Indicate the number of significant figures in each measured number below.
   (a) 520 mL (b) 3.0400 g (c) 0.005 00 m (d) 63,000 feet
6. Indicate the number of zeros that are significant in each measured number below.
   (a) 0.000 750 0 (b) 350.000 (c) 3000 (d) 0.000 75
7. Convert the following numbers from scientific notation to normal notation:
   (a) $7.45 \times 10^{+6}$ (b) 8.12×10^{4} (c) 3.41×10^{-1} (d) 7.3×10^{8}
8. Convert the following numbers from normal notation to scientific notation:
   (a) 0.000 784 2 (b) 86,000,000 (c) 0.123 (d) 58.3

9. Perform the following arithmetic operations and express the answer to the correct number of significant figures:

   (a) $32.991 + 45.3$ (b) $56.87 + 3.1$ (c) $0.005\,69 - 0.0003$

10. Perform the following arithmetic operations and express the answer to the correct number of significant figures:

    (a) $(8.32 \times 2.0) + 6.23$ (b) $\dfrac{3.6 \times 10^8}{82 - 4.2}$ (c) $\dfrac{3.90 \times 8.631}{6.0}$

11. Use unit analysis to perform each of the following conversions:

    (a) 927 L to mL (b) 0.564 m to cm (c) 563 cm^2 to m^2

12. Ethanol has a density of 0.789 g/mL. Calculate the volume occupied by 500.0 g of ethanol.

13. Gold has a density of 19.3 g/mL. What is the mass of a sample of gold that occupies a volume of 25.0 mL?

14. Calculate the heat (in joules) required to raise the temperature of 1000.0 g of water from 35.0°C to 60.0°C. (The specific heat of water is 4.18 J/g · C°.)

15. A sample of iron experiences a temperature increase of 10.0 C° when 2867 joules of heat are added to it. What is the mass of the iron? (The specific heat of iron is 0.449 J/g · C°.)

16. The speed limit on many city streets is 30 mi/h. What is this speed in km/h?

17. Convert 100.0°C to °F and express the answer to the correct number of significant figures.

18. Glucose has the formula $C_6H_{12}O_6$. Calculate the mass percents of the elements in glucose.

19. Calculate the weighted average atomic mass of naturally occurring copper, which is composed of two isotopes. The atomic number of Cu is 29, and the masses and percent abundances of the two isotopes are as follows:

    ^{63}Cu 69.17% isotope mass = 62.94

    ^{65}Cu 30.83% isotope mass = 64.93

20. Arrange the following elements in order of increasing ionization energy:

    Sb, I, Sn, Xe, Te

CHAPTER

4

The Modern Model of the Atom

Overview: What You Should Be Able to Do

Chapter 4 provides a model of atomic structure and describes the wave properties of electrons. The arrangement of electrons in atoms is presented, and a distinction is made between the ground state and the excited state arrangement of electrons in atoms. After mastering Chapter 4, you should be able to solve the following types of problems:

1. Calculate the energy when given the wavelength (or the wavelength when given the energy) of electromagnetic radiation.
2. Calculate the scaled energy absorbed or released by a hydrogen atom when an electron "jumps" to a higher energy shell or "falls" to a lower energy shell.
3. Write the electron configuration showing the electron distribution in an atom.
4. Write the shorthand electron configuration for an atom.

Chart 4.1 Calculating the energy when given the wavelength (or the wavelength when given the energy) of electromagnetic radiation

Step 1

- Determine the property that is to be calculated (i.e., energy or wavelength).

⬇

Step 2

- Use the relationship $E = \frac{hc}{\lambda}$ to find the value sought.

- If solving for energy, simply substitute the values h, c, and λ into the equation. (λ must be expressed in meters. Convert any other length unit to meters before substituting into the equation.)
- If solving for wavelength, algebraically manipulate the equation to isolate λ on the left side of the equation.
- The new equation becomes

$$\lambda = \frac{hc}{E}$$

- $h = 6.626 \times 10^{-34}\ \text{J} \cdot \text{s}$ (Planck's constant)
- $c = 3.00 \times 10^{8}\ \text{m/s}$ (speed of light)

⬇

Step 3

- Carry out the calculation.

Example 1

Calculate the energy of light that has a wavelength of 600 nm.

Solution

Perform step 1:

The quantity to be calculated is energy.

Perform step 2:

$$E = \frac{hc}{\lambda} = \frac{(6.626 \times 10^{-34}\ \text{J} \cdot \text{s})(3.00 \times 10^{8}\ \text{m/s})}{6.00 \times 10^{-7}\ \text{m}}$$

Perform step 3:

$$E = \frac{hc}{\lambda} = \frac{(6.626 \times 10^{-34}\ \text{J} \cdot \text{s})(3.00 \times 10^{8}\ \text{m/s})}{6.00 \times 10^{-7}\ \text{m}} = 3.31 \times 10^{-19}\ \text{J}$$

Example 2

What is the wavelength of electromagnetic radiation that has an energy of 4.42×10^{-19} J?

Solution

Perform step 1:

The quantity to be calculated is wavelength.

Perform step 2:

$$E = \frac{hc}{\lambda}$$

Algebraically manipulating the equation to solve for λ, we obtain

$$\lambda = \frac{hc}{E}$$

$$\lambda = \frac{hc}{E} = \frac{(6.626 \times 10^{-34}\ \text{J} \cdot \text{s})(3.00 \times 10^{8}\ \text{m/s})}{4.42 \times 10^{-19}\ \text{J}}$$

Perform step 3:

$$\lambda = \frac{hc}{E} = \frac{(6.626 \times 10^{-34}\ \text{J} \cdot \text{s})(3.00 \times 10^{8}\ \text{m/s})}{4.42 \times 10^{-19}\ \text{J}} = 4.50 \times 10^{-7}\ \text{m} = 450\ \text{nm}$$

Example 3

Calculate the energy of electromagnetic radiation that has a wavelength of 0.0100 m.

Solution

Perform step 1:

The quantity to be calculated is energy.

Perform step 2:

$$E = \frac{hc}{\lambda}$$

$$E = \frac{hc}{\lambda} = \frac{(6.626 \times 10^{-34}\ \text{J} \cdot \text{s})(3.00 \times 10^{8}\ \text{m/s})}{0.0100\ \text{m}}$$

Perform step 3:

$$E = \frac{hc}{\lambda} = \frac{(6.626 \times 10^{-34}\ \text{J} \cdot \text{s})(3.00 \times 10^{8}\ \text{m/s})}{0.0100\ \text{m}} = 1.99 \times 10^{-23}\ \text{J}$$

Practice Problems

4.1 Calculate the energy of electromagnetic radiation that has a wavelength of 350 nm.
4.2 What is the wavelength of electromagnetic radiation that has an energy of 4.09×10^{-19} J?
4.3 Green light has a wavelength of 486 nm. Calculate the energy of green light.

Chart 4.2 Calculating the scaled energy absorbed or released by a hydrogen atom when an electron "jumps" to a higher energy shell or "falls" to a lower energy shell

Step 1

- Determine the energy of the two shells involved in the electron transition. (Use the table on page 132 of the textbook to determine the energy values.)

⬇

Step 2

- Subtract the smaller energy value from the larger energy value.
- If the electron "jumps" from the lower energy shell to the higher energy shell, energy will be absorbed.
- If the electron "falls" from a higher energy shell to a lower energy shell, energy will be released.

Example 1

Calculate the amount of scaled energy absorbed when an electron in the $n = 1$ shell of the hydrogen atom jumps to the $n = 4$ shell.

Solution

Perform step 1:

Scaled energy of the $n = 1$ shell = 1.0 eV

Scaled energy of the $n = 4$ shell = 13.8 eV

Perform step 2:

The difference in energy is 13.8 eV − 1.0 eV = 12.8 eV.

Therefore 12.8 eV of scaled energy will be absorbed.

Example 2

Calculate the amount of scaled energy released when an electron in the $n = 5$ shell of the hydrogen atom falls to the $n = 3$ shell.

Solution

Perform step 1:

Scaled energy of the $n = 5$ shell = 14.1 eV

Scaled energy of the $n = 3$ shell = 13.1 eV

Perform step 2:

The difference in energy is 14.1 eV − 13.1 eV = 1.0 eV.
Therefore 1.0 eV of scaled energy will be released.

Example 3

Calculate the amount of scaled energy released when an electron in the $n = 3$ shell of the hydrogen atom falls to the $n = 1$ shell.

Solution

Perform step 1:

Scaled energy of the $n = 3$ shell = 13.1 eV
Scaled energy of the $n = 1$ shell = 1.0 eV

Perform step 2:

The difference in energy is 13.1 eV − 1.0 eV = 12.1 eV.
Therefore 12.1 eV of scaled energy will be released.

Practice Problems

4.4 Calculate the amount of scaled energy released when an electron in the $n = 6$ shell of the hydrogen atom falls to the $n = 2$ shell.

4.5 Calculate the amount of scaled energy released when an electron in the $n = 3$ shell of the hydrogen atom jumps to the $n = 5$ shell.

4.6 Calculate the amount of scaled energy released when an electron in the $n = 1$ shell of the hydrogen atom jumps to the $n = 6$ shell.

Chart 4.3 Writing the electron configuration showing the electron distribution in an atom

Step 1
- Determine the number of electrons in the atom.
- The number of electrons is equal to the atomic number for a neutral atom.

Step 2
- Place the electrons in subshells according to the periodic table shown on page 96 of the textbook.

Example 1

Write the electron configuration for silicon.

Solution

Perform step 1:

The atomic number of silicon is 14. There are 14 electrons in the silicon atom.

Perform step 2:

Following the order of placing electrons as demonstrated by the figure on page 140, we arrive at the following electron configuration for silicon.

$1s^2 2s^2 2p^6 3s^2 3p^2$

Example 2

Write the electron configuration for polonium.

Solution

Perform step 1:

The atomic number of polonium is 84. There are 84 electrons in the polonium atom.

Perform step 2:

Following the order of placing electrons as demonstrated by the figure on page 140, we arrive at the following electron configuration for polonium.

$1s^2 2s^2 2p^6 3s^2 3p^6 4s^2 3d^{10} 4p^6 5s^2 4d^{10} 5p^6 6s^2 4f^{14} 5d^{10} 6p^4$

Example 3

Write the electron configuration for zinc.

Solution

Perform step 1:

The atomic number of zinc is 30. There are 30 electrons in the zinc atom.

Perform step 2:

Following the order of placing electrons as demonstrated by the figure on page 140, we arrive at the following electron configuration for zinc.

$1s^2 2s^2 2p^6 3s^2 3p^6 4s^2 3d^{10}$

Practice Problems

4.7 Write the electron configuration for calcium.

4.8 Write the electron configuration for selenium.

4.9 Write the electron configuration for lead.

Chart 4.4 Writing the shorthand electron configuration for an atom

Step 1

- Determine the number of electrons in the atom.
- The number of electrons is equal to the atomic number for a neutral atom.

⬇

Step 2

- Identify the noble gas directly preceding the element whose configuration is to be written.
- Place the symbol of the noble gas in brackets.
- This symbol represents the electron configuration of the noble gas.

⬇

Step 3

- Write in the electrons following those in the noble gas whose symbol appears in the brackets.

Example 1

Using the noble gas abbreviated notation, write the electron configuration for germanium.

Solution

Perform step 1:

The atomic number of germanium is 32. There are 32 electrons in the germanium atom.

Perform step 2:

The noble gas immediately preceding germanium is argon (atomic number 18). Therefore the notation [Ar] represents the electron configuration through $3p^6$.

Perform step 3:

Writing the bracketed noble gas, and then following the order of placing electrons as demonstrated by the figure on page 140, we arrive at the following electron configuration for germanium.

$[Ar]4s^2 3d^{10} 4p^2$

Example 2

Using the noble gas abbreviated notation, write the electron configuration for barium.

Solution

Perform step 1:

The atomic number of barium is 56. There are 56 electrons in the barium atom.

Perform step 2:

The noble gas immediately preceding barium is xenon (atomic number 54). Therefore the notation [Xe] represents the electron configuration through $5p^6$.

Perform step 3:

Writing the bracketed noble gas, and then following the order of placing electrons as demonstrated by the figure on page 140, we arrive at the following electron configuration for germanium.

$[Xe]6s^2$

Example 3

Using the noble gas abbreviated notation, write the electron configuration for tellurium.

Solution

Perform step 1:

The atomic number of tellurium is 52. There are 52 electrons in the tellurium atom.

Perform step 2:

The noble gas immediately preceding tellurium is krypton (atomic number 36). Therefore the notation [Kr] represents the electron configuration through $4p^6$.

Perform step 3:

Writing the bracketed noble gas, and then following the order of placing electrons as demonstrated by the figure on page 140, we arrive at the following electron configuration for germanium.

$[Kr]5s^2 4d^{10} 5p^4$

Practice Problems

4.10 Write the electron configuration for magnesium, using the noble gas abbreviated notation.

4.11 Write the electron configuration for bismuth, using the noble gas abbreviated configuration.

4.12 Write the electron configuration for vanadium, using the noble gas abbreviated configuration.

Quiz for Chapter 4 Problems

1. How much scaled energy is absorbed when an electron in a hydrogen atom jumps from the $n = 3$ shell to the $n = 6$ shell?
2. Calculate the energy of electromagnetic radiation that has a wavelength of 400.0 nm.
3. How many electrons are in the *s* subshells in an atom of potassium?
4. What noble gas will appear in brackets in the noble gas abbreviated electron configuration for iodine?
5. How much scaled energy is released when an electron in a hydrogen atom falls from the $n = 4$ shell to the $n = 1$ shell?
6. Red light has a wavelength of 750 nm. What is the energy of red light?
7. Write the electron configuration for the phosphorus atom.
8. How does the energy released when an electron falls from the $n = 6$ shell to the $n = 5$ shell compare to the amount of energy absorbed when an electron jumps from the $n = 5$ shell to the $n = 6$ shell in the hydrogen atom?
9. Write the electron configuration for cadmium, using the noble gas abbreviated configuration.
10. What is the wavelength of electromagnetic radiation that has an energy of 3.50×10^{-18} joules?

CHAPTER

5

Chemical Bonding and Nomenclature

Overview: What You Should Be Able to Do

Chapter 5 provides a complete discussion of molecular and ionic bonding and presents the system for naming compounds. After mastering Chapter 5, you should be able to solve the following types of problems:

1. Draw dot diagrams for molecules and polyatomic ions.
2. Name binary ionic compounds in which the metal has the same ionic charge in all compounds.
3. Name binary ionic compounds in which the metal has different ionic charges in different compounds (most transition metals and some heavy representative metals).
4. Name binary covalent compounds.
5. Name binary ionic compounds that contain a polyatomic ion.
6. Name acids that are dissolved in water. (Acids that are not dissolved in water are named using the rules for naming binary covalent compounds or compounds containing polyatomic ions.)

Chart 5.1 Drawing dot diagrams for molecules and polyatomic ions

Step 1

- Determine the total number of valence electrons (dots) that will be in the diagram. Recall that the number of valence electrons is equal to the group number of the representative elements (e.g., N is in group 5A and therefore has five valence electrons).

- For a polyatomic ion, adjust the electron count to reflect the charge (for cations subtract the number of electrons equal to the charge; for anions add the number of electrons equal to the absolute value of the charge).

⬇

Step 2

- Connect the atoms using single covalent bonds.
- If the arrangement is not known, use the first nonhydrogen atom in the formula as the central atom.
- Add up the total number of electrons used in the single bonds, and subtract this number from the total number of electrons (calculated in step 1). This tells you the remaining electrons that must be placed as lone pairs in the dot diagram.

⬇

Step 3

- Add the remaining electrons as lone pairs, making sure that you satisfy the octet rule for each atom. Place the lone pairs on the central atom last.

⬇

Step 4

- Check to see if there are any atoms lacking an octet. If not, the dot structure is completed.
- If there are atoms lacking an octet, move lone pairs from adjacent atoms into bonding positions on the octet-deficient atom(s).

⬇

Step 5 (for polyatomic ions only)

- Place the entire dot structure in brackets, with the charge in the upper right-hand corner.

Example 1

Draw the electron dot diagram for H_2S.

Solution

Perform step 1:

The total number of valence electrons in H_2S is 8, as determined below.

H: 2×1 valence electron per H = 2 electrons

S: 1×6 valence electrons per S = 6 electrons

Total valence electrons = $2 + 6 = 8$

Perform step 2:

Connecting the atoms (with S as the central atom) using single bonds we obtain:

```
H:S:H
```

There are now four electrons in the structure, and four electrons remaining to be put in.

Perform step 3:

Placing the remaining electrons as lone pairs on the sulfur (remember that H never has lone pairs) we get the following dot structure for H_2S:

```
  ..
H:S:H
  ..
```

Example 2

Draw the electron dot diagram for SO_3^{2-}.

Solution

Perform step 1:

The total number of valence electrons in SO_3^{2-} is 26, as determined below.

S: 1×6 valence electrons per S = 6 electrons

O: 3×6 valence electrons per O = 18 electrons

2− charge: add 2 electrons = 2 electrons

Total electrons = $6 + 18 + 2 = 26$ electrons

Perform step 2:

Connecting the atoms (with S as the central atom) using single bonds we obtain

```
O:S:O
  ..
  O
```

There are now 6 electrons in the structure, leaving 20 electrons to be placed as lone pairs.

Perform step 3:

Placing the remaining 20 electrons as lone pairs on the atoms (leaving the S for last) and satisfying the octet rule we get the following dot structure:

```
 .. .. ..
:O:S:O:
 .. .. ..
   :O:
    ..
```

Perform step 4:

There are no atoms lacking an octet, so the dot structure is complete.

Perform step 5:

Because this is a polyatomic ion, the entire structure must be placed in brackets, with the 2− charge in the upper right-hand corner, as shown below.

$$\left[\begin{array}{c}:\!\ddot{\underset{\cdot\cdot}{O}}\!:\!\ddot{\underset{\cdot\cdot}{S}}\!:\!\ddot{\underset{\cdot\cdot}{O}}\!: \\ :\!\underset{\cdot\cdot}{O}\!:\end{array}\right]^{2-}$$

Example 3

Draw the electron dot diagram for N_2H_2.

Solution

Perform step 1:

The total number of valence electrons in N_2H_2 is 12, as determined below.

N: 2 × 5 valence electrons per N = 10 electrons
H: 2 × 1 valence electron per H = 2 electrons
Total electrons = 10 + 2 = 12 electrons

Perform step 2:

Connecting the atoms (with the N's as the central atoms) using single bonds we obtain

$$\mathrm{H{:}N{:}N{:}H}$$

There are now 6 electrons in the structure, leaving 6 electrons to be placed as lone pairs.

Perform step 3:

Placing the remaining 6 electrons as lone pairs on the atoms (until we run out of electrons) and satisfying the octet rule we get the following dot structure.

$$\mathrm{H{:}\ddot{\underset{\cdot\cdot}{N}}{:}\ddot{N}{:}H}$$

Perform step 4:

The nitrogen atom on the right is lacking an octet, so we move one of the lone pairs on the nitrogen on the left into a bonding position between the two nitrogens to obtain the dot structure

$$\mathrm{H{:}\ddot{N}{::}\ddot{N}{:}H}$$

Practice Problems

5.1 Draw the electron dot diagram for H_2O_2.

5.2 Draw the electron dot diagram for C_2H_4.

5.3 Draw the electron dot diagram for ClO_3^-.

Chart 5.2 Naming binary ionic compounds when the metal has the same ionic charge in all compounds

Step 1

- Name the metal ion (the positively charged cation) using the name of the metal.

⬇

Step 2

- Name the nonmetal ion (the negatively charged anion) by giving it an *-ide* ending.

⬇

Step 3

- Combine the name of the metal ion with that of the nonmetal ion to get the name of the compound.

Example 1

Name the compound Na_2O.

Solution

Perform step 1:

The name of the metal ion is sodium.

Perform step 2:

The name of the nonmetal ion is oxide.

Perform step 3:

The name of the compound is sodium oxide.

Example 2

Name the compound CaS.

Solution

Perform step 1:

The name of the metal ion is calcium.

Perform step 2:

The name of the nonmetal ion is sulfide.

Perform step 3:

The name of the compound is calcium sulfide.

Example 3

Name the compound Mg_3P_2.

Solution

Perform step 1:

The name of the metal ion is magnesium.

Perform step 2:

The name of the nonmetal ion is phosphide.

Perform step 3:

The name of the compound is magnesium phosphide.

Practice Problems

5.4 Name the compound $BaCl_2$.

5.5 Name the compound Ce_3N.

5.6 Name the compound KI.

Chart 5.3 Naming binary ionic compounds when the metal has different ionic charges in different compounds (most transition metals and some heavy representative metals)

Step 1

- Name the metal ion (the positively charged cation) using the name of the metal.

⬇

Step 2

- Specify the charge on the metal ion by placing a roman numeral indicating the charge in parentheses after the metal name, if giving the systematic name.
- Indicate the charge on the metal ion by using the appropriate suffix, if giving the old name.

⬇

Step 3

- Name the nonmetal ion (the negatively charged anion) by giving it an *-ide* ending.

⬇

Step 4

- Combine the name of the metal ion, the charge (if giving the systematic name), and the name of the nonmetal ion to get the name of the compound.

Example 1

Name the compound $CuBr_2$, giving the systematic name.

Solution

Perform step 1:

The name of the metal ion is copper.

Perform step 2:

The charge on the copper ion is +2, which must be indicated by the roman numeral II placed in parentheses. The name of the cation thus becomes copper(II).

Perform step 3:

The name of the nonmetal ion is bromide.

Perform step 4:

The name of the compound is copper(II) bromide.

Example 2

Name the compound $CuBr_2$, giving the old name.

Solution

Perform step 1:

The name of the metal ion is copper.

Perform step 2:

The charge on the copper ion is +2, which must be indicated by the using the suffix *-ic*. The name of the cation thus becomes cupric.

Perform step 3:

The name of the nonmetal ion is bromide.

Perform step 4:

The name of the compound is cupric bromide.

Example 3

Name the compound FeO, giving the systematic name.

Solution

Perform step 1:

The name of the metal ion is iron.

Perform step 2:

The charge on the iron ion is +2, which must be indicated by the roman numeral II placed in parentheses. The name of the cation thus becomes iron(II).

Perform step 3:

The name of the nonmetal ion is oxide.

Perform step 4:

The name of the compound is iron(II) oxide.

Practice Problems

5.7 Name the compound CoI_2, giving the systematic name.

5.8 Name the compound CuF, giving the old name.

5.9 Name the compound $HgBr_2$, giving the systematic name.

Chart 5.4 Naming binary covalent compounds

Step 1

- Name the less electronegative element first (this is usually the first element written in the formula).
- The element name is preceded by a Greek prefix (from Table 5.4 in the textbook) indicating the number of atoms of the element.
- If there is only one atom of the first element, the prefix *mono-* is not used.

⬇

Step 2

- Name the more electronegative element second, giving it the suffix *-ide*.
- The name is preceded by a Greek prefix indicating the number of atoms of the element.

⬇

Step 3

- Combine the names of both elements to get the name of the compound.

Example 1

Name the compound P_2O_5.

Solution

Perform step 1:

The name of the less electronegative element is phosphorus. There are two phosphorus atoms so the prefix *di-* is used, giving the name diphosphorus.

Perform step 2:

The name of the more electronegative element, with the suffix *-ide*, is oxide. There are five oxygen atoms present so the prefix *penta-* is used, giving the name pentoxide. (Note that the *a* is dropped from *penta-* to avoid having two vowels together.)

Perform step 3:

The name of the compound is diphosphorus pentoxide.

Example 2

Name the compound B_2H_6.

Solution

Perform step 1:

The name of the less electronegative element is boron. There are two boron atoms so the prefix *di-* is used. The name is diboron.

Perform step 2:

The name of the more electronegative element, with the suffix *-ide*, is hydride. There are six hydrogen atoms present so the prefix *hexa-* is used, giving the name hexahydride.

Perform step 3:

The name of the compound is diboron hexahydride.

Example 3

Name the compound OF_2.

Solution

Perform step 1:

The name of the less electronegative element is oxygen. There is only one oxygen atom so the prefix *mono-* is not used.

Perform step 2:

The name of the more electronegative element, with the suffix *-ide*, is fluoride. There are two fluoride atoms present so the prefix *di-* is used, giving the name difluoride.

Perform step 3:

The name of the compound is oxygen difluoride.

Practice Problems

5.10 Name the compound B_2O_3.

5.11 Name the compound NF_3.

5.12 Name the compound XeF_4.

Chart 5.5 Naming binary ionic compounds that contain a polyatomic ion

Step 1

- Name the metal ion (the positively charged cation) using the name of the metal.
- Specify the charge on the metal ion if the metal has different ionic charges in different compounds.
- The positive ion NH_4^+ is named ammonium.

⬇

Step 2

- Name the polyatomic anion, using the names given in Table 5.5 of the textbook.

⬇

Step 3

- Combine the name of the metal ion, the charge (if giving the systematic name), and the name of the polyatomic ion to get the name of the compound.

Example 1

Name the compound Li_3PO_4.

Solution

Perform step 1:

The name of the metal ion is lithium.

Perform step 2:

The name of the polyatomic ion is phosphate.

Perform step 3:

The name of the compound is lithium phosphate.

Example 2

Name the compound $Ca(NO_3)_2$.

Solution

Perform step 1:

The name of the metal ion is calcium.

Perform step 2:

The name of the polyatomic ion is nitrate.

Perform step 3:

The name of the compound is calcium nitrate.

Example 3

Name the compound $Fe(NO_2)_2$, giving the systematic name.

Solution

Perform step 1:

The name of the metal ion is iron.

Perform step 2:

The charge on the iron ion is +2, which must be indicated by the roman numeral II placed in parentheses. The name of the cation thus becomes iron(II).

Perform step 3:

The name of the polyatomic ion is nitrite.

Perform step 4:

The name of the compound is iron(II) nitrite.

Practice Problems

5.13 Name the compound TiF_3, giving the systematic name.
5.14 Name the compound NiO, giving the old name.
5.15 Name the compound $SnCl_4$, giving the systematic name.

Chart 5.6 Naming acids that are dissolved in water

Step 1

- Determine whether the acid contains oxygen.
- If no oxygen is present proceed to step 2.
- If oxygen is present proceed to step 3.

⬇

Step 2

- If the formula does not contain oxygen, add the prefix *hydro-* and the suffix *-ic acid* to the name of the nonhydrogen element or the polyatomic ion to get the name of the acid.

⬇

Step 3

- If the formula contains oxygen, name the acid based on the name of the polyatomic ion the acid contains.
- If the polyatomic ion name ends in *-ite,* replace the suffix *-ite* with *-ous acid* to name the acid.
- If the polyatomic ion name ends in *-ate,* replace the suffix *-ate* with *-ic acid* to name the acid.

Example 1

Name the acid HI.

Solution

Perform step 1:

The acid contains no oxygen; proceed to step 2.

Perform step 2:

Adding the prefix *hydro-* and the suffix *-ic acid* to the iodine gives the name hydroiodic acid.

Example 2

Name the acid $HClO_4$.

Solution

Perform step 1:

The acid contains oxygen; proceed to step 3.

Perform step 3:

The polyatomic ion ends in *-ate* (perchlorate) so we replace the suffix *-ate* with *-ic acid*, giving us the name perchloric acid.

Example 3

Name the acid HNO_3.

Solution

Perform step 1:

The acid contains oxygen; proceed to step 3.

Perform step 3:

The polyatomic ion ends in *-ate* (nitrate) so we replace the suffix *-ate* with *-ic acid*, giving us the name nitric acid.

Practice Problems

5.16 Name the acid $HClO_2$.

5.17 Name the acid $HBrO_4$.

5.18 Name the acid HF.

Quiz for Chapter 5 Problems

1. Draw electron dot diagrams for the following molecules:

   (a) PCl_3 (b) SO_2

2. Draw electron dot diagrams for the following polyatomic ions:

   (a) NH_4^+ (b) NO_2^-

3. Name the following compounds:

   (a) Ca_3P_2 (b) FeI_3

4. Name the following acids:

   (a) H_2SO_4 (b) HNO_2

5. Fill in the blanks in the following sentence:

   The charge on the copper in the compound CuO is __________, and the charge on the copper in the compound Cu_2O is __________.

6. Change each of the following incorrect names to correct names:

   (a) trimagnesium dinitride for Mg_3N_2

   (b) nitric acid for HNO_2

   (c) chlorous acid for HClO

7. Indicate the number of valence electrons that will appear in the dot structure for each species below.

   (a) $HClO_4$ (b) SO_4^{2-} (c) NH_4^+ (d) O_3

8. Which of the following atoms will never satisfy the octet rule in electron dot structures?

   (a) oxygen (b) boron (c) chlorine (d) hydrogen (e) sulfur

9. In which of the compounds below is it necessary to specify the charge on the metal ion when naming the compound? More than one answer is possible.

   (a) $BaCl_2$ (b) $CoCl_2$ (c) $BeCl_2$ (d) $HgCl_2$ (e) $CrCl_2$

10. Explain why it would be impossible to write the formula of a compound whose name is given as nitrogen oxide.

CHAPTER

6

The Shape of Molecules

Overview: What You Should Be Able to Do

Chapter 6 provides the procedures for determining the geometry and polarity of molecules, and discusses how the intermolecular attractive forces between molecules are dependent on molecular geometries. After mastering Chapter 6, you should be able to solve the following types of problems:

1. Use valence shell electron pair repulsion (VSEPR) theory to determine the shape and bond angles of molecules and polyatomic ions.
2. Use molecular shape to determine the polarity of molecules.

Chart 6.1 Using valence shell electron pair repulsion (VSEPR) theory to determine the shape and bond angles of molecules and polyatomic ions

Step 1

- Draw the correct electron dot structure for the molecule.

⬇

Step 2

- Count the number of electron groups (bonding groups plus lone pairs) around the central atom.
- Count each multiple bond as a single electron group.

⬇

Step 3

- Determine the electron-group geometry around the central atom, using the number of electron groups from step 2 and Table 6.2 (VSEPR Molecular Shape Table) in the textbook.

⬇

Step 4

- Pretend the lone pairs are invisible (if there are any lone pairs), and describe the resulting shape of the molecule.

⬇

Step 5

- Use the VSEPR Molecular Shape Table to determine the bond angles in the molecule.

Example 1

Determine the molecular shape and bond angles for $SiBr_4$.

Solution

Perform step 1:

The correct electron dot structure is

```
        ..
       :Br:
   ..   ..  ..
  :Br : Si : Br:
   ..   ..  ..
       :Br:
        ..
```

Perform step 2:

There are four groups of electrons around the central atom, Si.

Perform step 3:

According to the VSEPR Molecular Shape Table, a molecule with four groups of electrons around the central atom has a tetrahedral electron-group geometry.

Perform step 4:

There are no lone pairs on the central atom, so the molecular shape is also tetrahedral.

Perform step 5:

The bond angles for a tetrahedral molecule are 109.5°.

Example 2

Determine the molecular shape and bond angles for NF_3.

Solution

Perform step 1:

The correct electron dot structure is

```
 ··  ··  ··
:F :N: F:
 ··  ··  ··
    :F:
     ··
```

Perform step 2:

There are four groups of electrons around the central atom, N.

Perform step 3:

According to the VSEPR Molecular Shape Table, a molecule with four groups of electrons around the central atom has a tetrahedral electron-group geometry.

Perform step 4:

There is a lone pair on the central atom, so the molecular shape is pyramidal.

Perform step 5:

The bond angles for a pyramidal molecule are approximately 107°.

Example 3

Determine the molecular shape and bond angles for SO_3.

Solution

Perform step 1:

The correct electron dot structure is

```
 ··       ··
:O: S ::O
 ··  ··   ··
    :O:
     ··
```

Perform step 2:

There are three groups of electrons around the central atom, S. (The double bond counts as one electron group.)

Perform step 3:

According to the VSEPR Molecular Shape Table, a molecule with three groups of electrons around the central atom has a trigonal planar electron-group geometry.

Perform step 4:

There are no lone pairs on the central atom, so the molecular shape is trigonal planar.

Perform step 5:

The bond angles for a trigonal planar molecule are 120°.

Example 4

Determine the molecular shape and bond angles for OCS (C is the central atom).

Solution

Perform step 1:

The correct electron dot structure is:

:Ö::C::S̈:

Perform step 2:

There are two groups of electrons around the central atom, C. (Each double bond is counted as a single group.)

Perform step 3:

According to the VSEPR Molecular Shape Table, a molecule with two groups of electrons around the central atom has a linear electron-group geometry.

Perform step 4:

There are no lone pairs on the central atom, so the molecular shape is also linear.

Perform step 5:

The bond angles for a linear molecule are 180°.

Practice Problems

6.1 Determine the molecular shape and bond angles for Cl_2O.

6.2 Determine the molecular shape and bond angles for CO_3^{2-}.

6.3 Determine the molecular shape and bond angles for chloroform, $CHCl_3$.

6.4 Determine the molecular shape and bond angles for CS_2.

Chart 6.2 Using molecular shape to determine the polarity of molecules

Step 1

- Determine the molecular shape from the dot structure by applying VSEPR theory.

⬇

Step 2

- Draw vectors representing all bond dipole moments (using electronegativities).
- Determine the molecular dipole moment, if any.

Example 1

Determine whether OCS is a polar or a nonpolar molecule. Indicate the direction of the molecular dipole moment if one is present.

Solution

Perform step 1:

The correct electron dot structure is

:Ö::C::S̈:

VSEPR theory indicates that the molecule is linear because there are two electron groups around the central atom.

Perform step 2:

The individual bond dipoles are in the direction of the oxygen and the sulfur because O and S are more electronegative than carbon. The C–O bond dipole moment is larger because oxygen is more electronegative than sulfur.

$\longleftarrow\!\!\!+ \quad +\!\!\!\longrightarrow$

:Ö::C::S̈:

The molecular dipole is in the direction of the oxygen because the C–O bond dipole moment is larger than the C–S bond dipole moment.

Example 2

Determine whether CF_4 is a polar or a nonpolar molecule. Indicate the direction of the molecular dipole moment if one is present.

Solution

Perform step 1:

The correct electron dot structure is

```
    ..
   :F:
 .. .. ..
:F:C:F:
 .. .. ..
   :F:
    ..
```

VSEPR theory indicates that the molecule is tetrahedral because there are four electron groups around the central atom.

Perform step 2:

The bond dipoles are

```
     F↑
  ←+ | +
  F—C
  F    F
```

The individual bond dipoles are in the direction of the fluorine because fluorine is more electronegative than carbon. There is no net dipole because the individual dipole moments cancel one another. Therefore the CF_4 molecule is nonpolar.

Example 3

Determine whether $BFCl_2$ is a polar or a nonpolar molecule. Indicate the direction of the molecular dipole moment if one is present.

Solution

Perform step 1:

The correct electron dot structure is

```
 ..     ..
:F : B :Cl:
 ..  ..  ..
    :Cl:
     ..
```

VSEPR theory indicates that the molecule is trigonal planar because there are three electron groups around the central atom.

Perform step 2:

The bond dipoles are

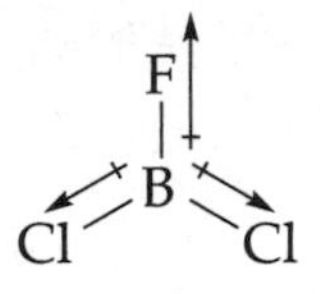

The individual bond dipoles are in the direction of the fluorine and the chlorines because fluorine and chlorine are more electronegative than boron. The net dipole is in the direction of the fluorine because fluorine is more electronegative than chlorine. Therefore the molecule is polar.

Practice Problems

6.5 Determine whether PI_3 is a polar or a nonpolar molecule. Indicate the direction of the molecular dipole moment if one is present.

6.6 Determine whether ONCl is a polar or a nonpolar molecule. Indicate the shape of the molecule and the direction of the molecular dipole moment if one is present.

6.7 Determine whether BCl_3 is a polar or a nonpolar molecule. Indicate the shape of the molecule and the direction of the molecular dipole moment if one is present.

Quiz for Chapter 6 Problems

1. Write the electron dot structure and indicate the molecular shape and bond angles for the HCOCl molecule. (The carbon atom is the central atom, and the three other atoms are connected to the carbon.)
2. Indicate the electron-group geometry for a molecule with the following number of electron groups around the central atom:

   (a) two electron groups

   (b) three electron groups

   (c) four electron groups
3. Predict the molecular shape and the bond angles for the ClO_3^- ion.
4. Use VSEPR theory to determine whether the following molecules are polar or nonpolar. If the molecule is polar, indicate the direction of the net dipole moment.

   (a) CH_4 (b) CH_3F (c) CH_2F_2 (d) CHF_3 (e) CF_4
5. Arrange the following bonds in order of increasing bond dipole moment.

   C—O C—F C—Br C—H C—Cl
6. Indicate whether the following statement is true or false, and explain why it is false if false.

   The molecular shape is tetrahedral for all molecules in which the central atom is surrounded by four electron groups.
7. The molecular shape of a molecule containing two electron groups and no lone pairs around the central atom is ____________.
8. Circle the nonpolar molecule(s) below.

   NF_3 BF_3 SO_3 PF_3
9. The electron-group geometry for Cl_2O is ________, and the molecular shape is ___________.
10. There are two different bent molecular shapes, one with bond angles of approximately 105° and the other with bond angles of approximately 118°. Indicate (a) the number of total electron groups, (b) the number of bonding pairs, and (c) the number of lone pairs that are associated with each of these two molecular shapes.

Cumulative Quiz for Chapters 4, 5, and 6

1. How much scaled energy is absorbed when an electron in a hydrogen atom jumps from the $n = 2$ shell to the $n = 4$ shell?
2. How many electrons are in the *s* subshells in an atom of bromine?
3. What is the energy of blue light that has a wavelength of 475 nm?
4. Write the ground-state electron configuration for a strontium atom.
5. What is the wavelength of electromagnetic radiation that has an energy of 3.50×10^{-18} joules?
6. Draw electron dot diagrams for the following molecules:

   (a) NBr_3 (b) BF_3
7. Draw electron dot diagrams for the following polyatomic ions:

   (a) CO_3^{2-} (b) NH_4^+
8. Name the following compounds:

   (a) Mg_3N_2 (b) CoS (c) $HClO_3$ (d) HNO_2
9. Indicate the number of valence electrons that will appear in the dot structure for each species below:

   (a) SO_3^{2-} (b) H_2CO_3 (c) P_4
10. Which of the following atom(s) do not satisfy the octet rule in electron dot structures of neutral compounds?

    (a) hydrogen (b) fluorine (c) boron (e) argon
11. Write the electron dot structure and indicate the molecular shape and bond angles for the CH_3Br molecule.
12. Predict the molecular geometry and the bond angles for the SO_3^{2-} ion.
13. Use VSEPR to determine whether the following molecules are polar or nonpolar. If the molecule is polar, indicate the direction of the net dipole moment.

    (a) BF_3 (b) CH_2Cl_2 (c) H_2S (d) BeH_2
14. The molecular shape of a molecule containing three electron groups and one lone pair around the central atom is ______________.

15. Circle the nonpolar molecular molecule(s) below.

PI_3 BI_3 CH_2F_2

16. The electron-group geometry for NO_2^- is __________, and the molecular shape is __________.

17. There are two different geometries for a molecule that has the general formula XY_3. What are these two shapes, and how do they differ in terms of the number of electron groups and lone pairs around the central atom?

18. Arrange the following bonds in order of increasing bond dipole moment:

H–O H–F H–Br H–H H–Cl

19. Determine whether CS_2 is a polar or nonpolar molecule. Indicate the direction of the molecular dipole moment if one is present.

20. Indicate the bond angles for each of the electron-group geometries below.

(a) trigonal planar (b) pyramidal (c) tetrahedral

CHAPTER

7

Chemical Reactions

Overview: What You Should Be Able to Do

Chapter 7 presents the various types of chemical reactions and provides a method for balancing chemical equations. After mastering Chapter 7, you should be able to solve the following types of problems:

1. Balance chemical equations by the inspection method.
2. Identify the precipitate that will form in a potential precipitation reaction, and write the net ionic equation for the reaction.
3. Write the equation for the neutralization reaction that will occur when an acid reacts with a base, and write the net ionic equation for the reaction.

Chart 7.1 Balancing chemical equations by the inspection method

Step 1

- Scan the equation from left to right, placing appropriate coefficients to make the number of atoms of each element the same on both sides of the equation.
- Never change the subscripts in a formula to balance an equation; this will change the identity of the substance.
- It is often helpful to save for last substances that appear as pure elements (e.g., H_2, N_2, O_2, Al).
- Fractional coefficients (e.g., $\frac{1}{2}$, $\frac{3}{2}$) can be used to balance the equation.

⬇

Step 2

- Convert all fractions to whole numbers by multiplying each coefficient by the denominator of the fraction.
- Check to make sure that each side of the equation contains the same number of atoms of each element.

Example 1

Balance the following equation:

$$C_5H_{12} + O_2 \longrightarrow CO_2 + H_2O$$

Solution

Perform step 1:

Scanning the reaction from left to right (leaving the O_2 for last), and adding coefficients to balance each element, we obtain

$$C_5H_{12} + O_2 \longrightarrow 5\ CO_2 + 6\ H_2O$$

Balancing the oxygens, we obtain

$$C_5H_{12} + 8\ O_2 \longrightarrow 5\ CO_2 + 6\ H_2O$$

Perform step 2:

Checking the number of atoms on each side of the equation, we get

C: 5 atoms on each side
H: 12 atoms on each side
O: 16 atoms on each side

Therefore the equation is balanced.

Example 2

Balance the following equation:

$$C_5H_{10}O_2 + O_2 \longrightarrow CO_2 + H_2O$$

Solution

Perform step 1:

Scanning the reaction from left to right (leaving the oxygens for last), and adding coefficients to balance each element, we obtain

$$C_5H_{10}O_2 + O_2 \longrightarrow 5\ CO_2 + 5\ H_2O$$

To balance the oxygens, we need an additional 13 O's on the left. We must use the coefficient $\frac{13}{2}$ to obtain 13 O's from O_2. The equation then becomes

$$C_5H_{10}O_2 + \frac{13}{2}\,O_2 \longrightarrow 5\,CO_2 + 5\,H_2O$$

Perform step 2:

The coefficients can be converted to whole numbers by multiplying all of them by 2 (to convert the $\frac{13}{2}$ to 13). The equation then becomes

$$2\,C_5H_{10}O_2 + 13\,O_2 \longrightarrow 10\,CO_2 + 10\,H_2O$$

Checking the number of atoms on each side of the equation, we get

C: 10 atoms on each side
H: 20 atoms on each side
O: 30 atoms on each side

Therefore the equation is balanced.

Example 3

Balance the following equation:

$$Mg_3N_2 + H_2O \longrightarrow NH_3 + Mg(OH)_2$$

Solution

Perform step 1:

Scanning the reaction from left to right, and adding coefficients to balance each element, we obtain

$$Mg_3N_2 + 6\,H_2O \longrightarrow 2\,NH_3 + 3\,Mg(OH)_2$$

Perform step 2:

Checking the number of atoms on each side of the equation, we get

Mg: 3 atoms on each side
N: 2 atoms on each side
H: 12 atoms on each side
O: 6 atoms on each side

Therefore the equation is balanced.

Practice Problems

7.1 Balance the following equation:

$$NCl_3 + H_2O \longrightarrow NH_3 + HOCl$$

7.2 Balance the following equation:

$$Al_2O_3 \longrightarrow Al + O_2$$

7.3 Balance the following equation:

$$PCl_5 + H_2O \longrightarrow H_3PO_4 + HCl$$

Chart 7.2 Identifying the precipitate that will form in a potential precipitation reaction, and writing the net ionic equation for the reaction

Step 1

- Identify all ions present when the starting solutions are mixed. (These are the ions produced when the starting materials dissolved.)

⬇

Step 2

- Identify all possible cation–anion combinations that are different from the ones in the starting materials. These combinations represent possible precipitates.

⬇

Step 3

- Determine any precipitates (insoluble salts) that will form from any of the cation–anion combinations. Table 7.1 in the textbook provides the solubility rules.

⬇

Step 4

- If a precipitation reaction occurs, write the balanced equation, the ionic equation, and the net ionic equation for it, making sure that the molecular formula for the precipitate is electrically neutral.

Example 1

A solution of sodium chloride is combined with a solution of silver nitrate. Write the balanced equation for the precipitation reaction that occurs, the ionic equation, and the net ionic equation.

Solution

Perform step 1:

The ions present in the solution will be Na^+, Cl^-, Ag^+, and NO_3^-.

Perform step 2:

The possible ion combinations to form a precipitate are Na^+ with NO_3^- and Ag^+ with Cl^-.

Perform step 3:

According to the solubility rules, only Ag^+ and Cl^- will form a precipitate.

Perform step 4:

The balanced equation for the reaction that occurs is

$$NaCl(aq) + AgNO_3(aq) \longrightarrow NaNO_3(aq) + AgCl(s)$$

The ionic equation for the reaction that occurs is

$$Na^+(aq) + Cl^-(aq) + Ag^+(aq) + NO_3^-(aq) \longrightarrow Na^+(aq) + NO_3^-(aq) + AgCl(s)$$

The net ionic equation is

$$Ag^+(aq) + Cl^-(aq) \longrightarrow AgCl(s)$$

Example 2

A solution of $Ca(NO_3)_2$ is combined with a solution of Na_2SO_4. Write the balanced equation for the precipitation reaction that occurs, the ionic equation, and the net ionic equation.

Solution

Perform step 1:

The ions present in the solution will be Ca^{2+}, NO_3^-, Na^+, and SO_4^{2-}.

Perform step 2:

The possible ion combinations to form a precipitate are Ca^{2+} with SO_2^{2-} to form $CaSO_4$, and Na^+ with NO_3^- to form $NaNO_3$.

Perform step 3:

According to the solubility rules, only $CaSO_4$ will form a precipitate.

Perform step 4:

The balanced equation for the precipitation reaction that occurs is

$$Ca(NO_3)_2(aq) + Na_2SO_4(aq) \longrightarrow CaSO_4(s) + 2\ NaNO_3(aq)$$

The ionic equation is

$$Ca^{2+}(aq) + 2\ NO_3^-(aq) + 2\ Na^+(aq) + SO_4^{2-}(aq) \longrightarrow \\ CaSO_4(s) + 2\ Na^+(aq) + 2\ NO_3^-(aq)$$

The net ionic equation is

$$Ca^{2+}(aq) + SO_4^{2-}(aq) \longrightarrow CaSO_4(s)$$

Practice Problems

7.4 A solution of NaOH is mixed with a solution of $MgCl_2$. Write the balanced equation for the precipitation reaction that occurs, the ionic equation, and the net ionic equation.

7.5 A solution of K_3PO_4 is mixed with a solution of $Al(NO_3)_3$. Write the balanced equation for the precipitation reaction that occurs, the ionic equation, and the net ionic equation.

Chart 7.3 Writing the equation for the neutralization reaction that will occur when an acid reacts with a base, and writing the net ionic equation for the reaction

Step 1

- Write the balanced intact molecule equation (the molecular equation) for the reaction.

⬇

Step 2

- Write the reactants for the reaction (remember the coefficients).
- Break the acid into its H^+'s and anion, and break the base into its OH^-'s and cation.

⬇

Step 3

- Write the products of the reaction.
- Combine the cation from the base with the anion from the acid to make a salt. (The number of cations and anions in the formula of the salt must be such that the salt is electrically neutral.)
- Combine the H^+ and the OH^- to form water.

⬇

Step 4

- Write the ionic neutralization reaction equation.
- Write the net ionic equation for the reaction, $H^+ + OH^- \longrightarrow H_2O$, by cancelling species that appear on both sides of the equation.

Example 1

Write the molecular neutralization reaction equation, the ionic equation, and the net ionic equation for the reaction that occurs between HI and KOH.

Solution

Perform step 1:

$HI(aq) + KOH(aq) \longrightarrow KI(aq) + H_2O(l)$

Perform step 2:

The reactants are $H^+(aq) + I^-(aq) + K^+(aq) + OH^-(aq)$.

Perform step 3:

The products are $K^+(aq) + I^-(aq) + H_2O(l)$.

Perform step 4:

The ionic equation for the neutralization reaction is

$$H^+(aq) + I^-(aq) + K^+(aq) + OH^-(aq) \longrightarrow K^+(aq) + I^-(aq) + H_2O(l)$$

The net ionic equation is $H^+(aq) + OH^-(aq) \longrightarrow H_2O(l)$.

Example 2

Write the molecular neutralization reaction equation, the ionic equation, and the net ionic equation for the reaction that occurs between HNO_3 and $Ca(OH)_2$.

Solution

Perform step 1:

$2\,HNO_3(aq) + Ca(OH)_2(aq) \longrightarrow Ca(NO_3)_2(aq) + 2\,H_2O(l)$

Perform step 2:

The reactants are $2\,H^+(aq) + 2\,NO_3^-(aq) + Ca^{2+}(aq) + 2\,OH^-(aq)$.

Perform step 3:

The products of the reaction are $Ca^{2+}(aq) + 2\,NO_3^-(aq) + 2\,H_2O(l)$.

Perform step 4:

The ionic equation for the neutralization reaction is

$$2\,H^+(aq) + 2\,NO_3^-(aq) + Ca^{2+}(aq) + 2\,OH^-(aq) \longrightarrow Ca^{2+}(aq) + 2\,NO_3^-(aq) + 2\,H_2O(l)$$

The net ionic equation is

$$2\,H^+(aq) + 2\,OH^-(aq) \longrightarrow 2\,H_2O(l)$$

which reduces to

$$H^+(aq) + OH^-(aq) \longrightarrow H_2O(l).$$

Practice Problems

7.6 Write the molecular neutralization reaction equation, the ionic equation, and the net ionic equation for the reaction that occurs between HBr and KOH.

7.7 Write the molecular neutralization reaction equation, the ionic equation, and the net ionic equation for the reaction that occurs between HNO_3 and $Ba(OH)_2$.

Quiz for Chapter 7 Problems

1. Balance the equation for the reaction

$$C_{10}H_{20} + O_2 \longrightarrow CO_2 + H_2O.$$

2. The coefficient in front of the PH_3 when the equation below is balanced is _____.

$$PH_3 + O_2 \longrightarrow P_4O_{12} + H_2O$$

3. Which of the following ionic salts is (are) insoluble in water?

   (a) $Ba(NO)_3$ (b) $MgCO_3$ (c) $CaSO_4$ (d) NH_4Cl

4. Write the formula of the precipitate that forms when a solution of K_2CO_3 is mixed with a solution of $CaBr_2$.
5. Write all of the ions that are present in solution when a solution of LiCl is mixed with a solution of $Ba(NO_3)_2$.
6. Write the molecular equation, the ionic equation, and the net ionic equation for the reaction that occurs when Na_2SO_4 reacts with $Pb(NO_3)_2$.
7. When an acid is mixed with a base, the products of the reaction are a(n) _________ and _________.
8. Write the molecular equation, the ionic equation, and the net ionic equation for the precipitation reaction that occurs when a solution of NH_4I is mixed with a solution of $Pb(NO_3)_2$.
9. Which of the following ionic salts is (are) insoluble in water?

   (a) Na_2CO_3 (b) $BaCl_2$ (c) $CaSO_4$ (d) NH_4NO_3

10. The reaction between an acid and a base is called a(n) ______________ reaction.
11. When balancing chemical equations, only the ____________ of the reactants and products can be changed because altering the ____________ causes a change in the chemical identity of a substance.

CHAPTER

8

Stoichiometry and the Mole

Overview: What You Should Be Able to Do

Chapter 8 introduces the quantitative aspects of chemical reactions and the procedures for performing stoichiometric calculations. After mastering Chapter 8, you should be able to solve the following types of problems:

1. Calculate theoretical yield of products from grams (or moles) of one or more reactants.
2. Solve limiting-reactant problems to determine the theoretical yield(s) when amounts of two reactants are given.
3. Determine the formula of a compound ($C_xH_yO_z$) from combustion-analysis data.
4. Convert an empirical formula to a molecular formula.
5. Determine the percent composition from the molecular formula.
6. Determine the empirical formula from the percent composition.

Chart 8.1 Calculating theoretical yield of products from grams (or moles) of one or more reactants

Step 1

- Write the balanced chemical equation for the reaction.

⬇

Step 2

- Convert the given information (grams, atoms, or molecules) for any reactants to moles.
- If moles are given, proceed to step 3.
- Use the molar mass to convert grams to moles.
- Use Avogadro's number to convert atoms or molecules to moles.

⬇

Step 3

- Use the coefficients from the balanced equation to find the relationship between the number of moles of the given substance and the number of moles of the unknown substance.
- The ratio of the coefficients is used as a conversion factor, with the coefficient in front of the unknown as the numerator and the coefficient in front of the given as the denominator.
- The conversion factor is multiplied by the number of moles of the given substance.

⬇

Step 4

- Convert the moles of the unknown, calculated in step 3, to grams (or atoms or molecules).

Example 1

Calculate the grams of hydrogen required to produce 82.00 grams of ammonia, NH_3, from N_2 and H_2.

Solution

Perform step 1:

The balanced equation is

$$N_2 + 3\,H_2 \longrightarrow 2\,NH_3$$

Perform step 2:

$$\text{Moles of } NH_3 = 82.00\text{ g} \times \frac{1\text{ mol } NH_3}{17.03\text{ g } NH_3} = 4.184\text{ mol } NH_3$$

Perform step 3:

$$\text{Moles of } H_2 = 4.184\text{ mol } NH_3 \times \frac{3\text{ mol } H_2}{2\text{ mol } NH_3} = 6.276\text{ mol } NH_3$$

Perform step 4:

$$\text{Grams of } H_2 = 6.276 \text{ mol } H_2 \times \frac{2.016 \text{ g } H_2}{1 \text{ mol } H_2} = 12.65 \text{ g } H_2$$

Example 2

In the reaction of O_2 with C_2H_4 to produce CO_2 and H_2O, how many moles of O_2 are needed to completely react with 25.0 g of C_2H_4? How many atoms of O_2 are needed?

Solution

Perform step 1:

The balanced equation is

$$C_2H_4 + 3\ O_2 \longrightarrow 2\ CO_2 + 2\ H_2O$$

Perform step 2:

$$\text{Moles of } C_2H_4 = 25.0 \text{ g} \times \frac{1 \text{ mol } C_2H_4}{28.05 \text{ g } C_2H_4} = 0.891 \text{ mol } C_2H_4$$

Perform step 3:

$$\text{Moles of } O_2 = 0.891 \text{ mol } C_2H_4 \times \frac{3 \text{ mol } O_2}{1 \text{ mol } C_2H_4} = 2.67 \text{ mol } O_2$$

Perform step 4 to determine number of atoms:

$$\text{Atoms of } O_2 = 2.67 \text{ mol } O_2 \times \frac{6.022 \times 10^{23} \text{ atoms } O_2}{1 \text{ mol } O_2} = 1.61 \times 10^{24} \text{ atoms } O_2$$

Practice Problems

8.1 Calculate the number of grams of NH_3 that can be produced from 500.0 g of NCl_3 according to the following (unbalanced) equation:

$$NCl_3 + H_2O \longrightarrow NH_3 + HOCl$$

8.2 Calculate the number of moles of NaOH that are needed to react with 500.0 g of H_2SO_4 according to the following (unbalanced) equation:

$$H_2SO_4 + NaOH \longrightarrow Na_2SO_4 + H_2O$$

Chart 8.2 Solving limiting-reactant problems to determine the theoretical yield when amounts of two reactants are given

Step 1

- Write the balanced chemical equation for the reaction.

⬇

Step 2

- Convert the given information (grams, atoms, or molecules) *for all reactants* to moles.
- If moles for all reactants are given, proceed to step 3.
- Use the molar mass to convert grams to moles.
- Use Avogadro's number to convert atoms or molecules to moles.

⬇

Step 3

- Determine the mole-to-coefficient ratio for each reactant, using the appropriate coefficients from the balanced equation.
- The reactant that has the smallest mole-to-coefficient ratio is the limiting reactant.

⬇

Step 4

- Use the coefficients from the balanced equation and the molar amount of the limiting reactant to calculate the theoretical yield of the product in moles.

⬇

Step 5

- Convert the moles of the unknown, calculated in step 3, to grams (or atoms or molecules).

Example 1

In the reaction of HCl with Zn to produce H_2 gas and $ZnCl_2$, how many grams of H_2 gas can be made from 90.0 grams of HCl with 50.0 grams of Zn?

Solution

Perform step 1:

The balanced equation for the reaction is

$$2\ HCl + Zn \longrightarrow H_2 + ZnCl_2$$

Perform step 2:

$$\text{Moles of HCl} = 90.0 \text{ g HCl} \times \frac{1 \text{ mol HCl}}{36.46 \text{ g HCl}} = 2.46 \text{ mol HCl}$$

$$\text{Moles of Zn} = 50.0 \text{ g Zn} \times \frac{1 \text{ mol Zn}}{65.39 \text{ g Zn}} = 0.765 \text{ mol Zn}$$

Perform step 3:

Mole-to-coefficient ratios:

$$\text{HCl: } \frac{2.46}{2} = 1.23 \qquad \text{Zn: } \frac{0.765}{1} = 0.765$$

Zn has the smaller mole-to-coefficient ratio and is therefore the limiting reactant.

Perform step 4:

$$\text{Moles of } H_2 = 0.765 \text{ mol Zn} \times \frac{1 \text{ mol } H_2}{2 \text{ mol HCl}} = 0.383 \text{ mol } H_2$$

Perform step 5:

$$\text{Grams of } H_2 = 0.383 \text{ mol } H_2 \times \frac{2.016 \text{ g } H_2}{1 \text{ mol } H_2} = 0.772 \text{ g } H_2$$

Example 2

In the production of NBr_3 from N_2 and Br_2 according to the unbalanced equation below, how many grams of NBr_3 can be made from 20.0 grams of N_2 with 300.0 grams of Br_2?

$$N_2 + Br_2 \longrightarrow NBr_3$$

Solution

Perform step 1:

The balanced equation for the reaction is $N_2 + 3\,Br_2 \longrightarrow 2\,NBr_3$.

Perform step 2:

$$\text{Moles of } N_2 = 20.0 \text{ g } N_2 \times \frac{1 \text{ mol } N_2}{28.014 \text{ g } N_2} = 0.714 \text{ mol } N_2$$

$$\text{Moles of } Br_2 = 300.0 \text{ g } Br_2 \times \frac{1 \text{ mol } Br_2}{159.8 \text{ g } Br_2} = 1.877 \text{ mol } Br_2$$

Perform step 3:

Mole-to-coefficient ratios:

$$N_2\text{: } \frac{0.714}{1} = 0.714 \qquad Br_2\text{: } \frac{1.877}{3} = 0.6258$$

Br_2 is the limiting reactant because it has the smaller mole-to-coefficient ratio.

Note that although there are many more grams of Br_2 than grams of N_2, when the moles of each substance and the coefficients in the balanced equation are considered, Br_2 is limiting.

Perform step 4:

$$\text{Moles of } NBr_3 = 1.877 \text{ mol } Br_2 \times \frac{2 \text{ mol } NBr_3}{3 \text{ mol } Br_2} = 1.251 \text{ mol } NBr_3$$

Perform step 5:

$$\text{Grams of } NBr_3 = 1.251 \text{ mol } NBr_3 \times \frac{253.7 \text{ g } NBr_3}{1 \text{ mol } NBr_3} = 317.1 \text{ g } NBr_3$$

Practice Problems

8.3 In the reaction of H_2 with O_2 to produce H_2O, according to the unbalanced equation below, how many grams of water can be produced from the reaction of 150.0 grams of H_2 with 200.0 grams of O_2?

$$H_2 + O_2 \longrightarrow H_2O$$

8.4 In the reaction of C_3H_8 with O_2 to produce CO_2 and H_2O, according to the unbalanced equation below, how many grams of CO_2 can be produced from the reaction of 25.0 grams of C_3H_8 with 75.00 grams of O_2?

$$C_3H_8 + O_2 \longrightarrow CO_2 + H_2O$$

Chart 8.3 Determining the formula of a compound ($C_xH_yO_z$) from combustion-analysis data

Step 1

- Convert grams of CO_2 to moles of C. Use the formula shown below to do the conversion.

$$\text{Moles of C} = \text{mol } CO_2 = \text{g } CO_2 \times \frac{1 \text{ mol } CO_2}{44.009 \text{ g } CO_2}$$

- Convert grams of H_2O to moles of H. Use the formula shown below to do the conversion.

$$\text{Moles of H} = 2 \times \text{mol } H_2O = 2 \times \text{g } H_2O \times \frac{1 \text{ mol } H_2O}{18.015 \text{ g } H_2O}$$

⬇

Step 2

- Convert moles of C from step 1 to grams of C.
- Convert moles of H from step 1 to grams of H.

⬇

Step 3

- Determine the mass of O, if any, by subtracting the sum of the grams of C + H from the grams of the sample burned.
- If this number is approximately zero (0.01 or less), the subscript z is zero and the compound contains no oxygen.
- If this number is not zero, convert it to moles of O.

⬇

Step 4

- Divide the value of all moles from steps 2 and 3 by the smallest one, and if the resulting values are very close to whole numbers, round them to whole numbers.
- These are the values of x, y, and z in the empirical formula $C_xH_yO_z$.

⬇

Step 5

- If one or more subscripts are far from a whole number, find some multiplier that makes them whole.
- Multiply all the subscripts by the multiplier that makes whole numbers.
- The multipliers for some values of subscripts that are far from whole numbers are given below.

Approximate decimal value	Multiplier
0.25, 0.75	4
0.33, 0.67	3
0.50	2
0.167	6
0.13, 0.38, 0.63, 0.88	8

Example 1

A compound containing C, H, and O is subjected to combustion analysis. 30.03 grams of compound produced 43.5 grams of CO_2 and 23.5 grams of H_2O. What is the empirical formula of the compound?

Solution

Perform step 1:

$$\text{Moles of C} = \text{moles of } CO_2 = 43.5 \text{ g } CO_2 \times \frac{1 \text{ mol } CO_2}{44.0098 \text{ g } CO_2} = 0.988 \text{ mol C}$$

$$\text{Moles of H} = 2 \times \text{moles of } H_2O = 2 \times 23.5\text{g } H_2O \times \frac{1 \text{ mol } H_2O}{18.015 \text{ g } H_2O} = 2.61 \text{ mol H}$$

Perform step 2:

$$\text{Grams of C} = 0.988 \text{ mol C} \times \frac{12.011 \text{ g C}}{1 \text{ mol C}} = 11.9 \text{ g C}$$

$$\text{Grams of H} = 2.61 \text{ mol H} \times \frac{1.007\,94 \text{ g H}}{1 \text{ mol H}} = 2.63 \text{ g H}$$

Perform step 3:

$$\text{Grams of O} = 30.03\text{-g sample} - (11.9 \text{ g C} + 2.63 \text{ g H}) = 15.5 \text{ g O}$$

$$\text{Moles of O} = 15.5 \text{ g O} \times \frac{1 \text{ mol O}}{15.9994 \text{ g O}} = 0.969 \text{ mol O}$$

Perform step 4:

The mole values are as follows: 0.988 mol C, 2.61 mol H, and 0.969 mol O. 0.969 mol O is the smallest value, so we divide each of the mole values by this number.

$$\text{Subscript for C: } \frac{0.988}{0.969} = 1.02$$

$$\text{Subscript for H: } \frac{2.61}{0.969} = 2.69$$

$$\text{Subscript for O: } \frac{0.969}{0.969} = 1.00$$

These are the values of x, y, and z. The empirical formula would be $C_{1.02}H_{2.69}O_{1.00}$.

Perform step 5 (necessary because 2.69 is far from a whole number):

Because 2.69 is approximately equal to 2.67 (which equals $2\frac{2}{3}$ or $\frac{8}{3}$), it can be multiplied by 3 to make it a whole number. We therefore multiply all of the subscripts by 3. The empirical formula is $C_3H_8O_3$.

Example 2

A compound containing C, H, and O is subjected to combustion analysis. 7.49 grams of compound produced 14.96 grams of CO_2 and 6.13 grams of H_2O. What is the empirical formula of the compound?

Solution

Perform step 1:

$$\text{Moles of C} = \text{moles of } CO_2 = 14.96 \text{ g } CO_2 \times \frac{1 \text{ mol } CO_2}{44.0098 \text{ g } CO_2} = 0.340 \text{ mol C}$$

$$\text{Moles of H} = 2 \times \text{moles of } H_2O = 2 \times 6.13 \text{ g } H_2O \times \frac{1 \text{ mol } H_2O}{18.0152 \text{ g } H_2O} = 0.681 \text{ mol H}$$

Perform step 2:

$$\text{Grams of C} = 0.340 \text{ mol C} \times \frac{12.011 \text{ g C}}{1 \text{ mol C}} = 4.08 \text{ g C}$$

$$\text{Grams of H} = 0.681 \text{ mol H} \times \frac{1.007\,94 \text{ g H}}{1 \text{ mol H}} = 0.686 \text{ g H}$$

Perform step 3:

$$\text{Grams of O} = 7.49\text{-g sample} - (4.08 \text{ g C} + 0.686 \text{ g H}) = 2.72 \text{ g O}$$

$$\text{Moles of O} = 2.72 \text{ g O} \times \frac{1 \text{ mol O}}{15.9994 \text{ g O}} = 0.170 \text{ mol O}$$

Perform step 4:

The mole values are as follows: 0.340 mol C, 0.681 mol H, and 0.170 mol O. 0.170 mol O is the smallest value, so we divide each of the mole values by this number.

$$\text{Subscript for C: } \frac{0.340}{0.170} = 2.00$$

$$\text{Subscript for H: } \frac{0.681}{0.170} = 4.01$$

$$\text{Subscript for O: } \frac{0.170}{0.170} = 1.00$$

These are the values of x, y, and z in the empirical formula $C_xH_yO_z$. The empirical formula is therefore C_2H_4O.

Practice Problems

8.5 When 1.65 grams of hydroquinone is subjected to combustion analysis, 3.96 grams of CO_2 and 0.81 grams of H_2O are produced. What is the empirical formula of hydroquinone?

8.6 0.762 grams of a compound yielded 2.39 grams of CO_2 and 0.9749 grams of H_2O. What is the empirical formula of the compound?

Chart 8.4 Converting an empirical formula to a molecular formula

Step 1

- Determine the molar mass of the empirical formula.

⬇

Step 2

- Divide the molar mass of the compound by the molar mass of the empirical formula.
- This will yield a number very close to a whole number, and should be rounded to a whole number.

⬇

Step 3

- Multiply all subscripts in the empirical formula by the number calculated in step 2.

Example 1

A compound whose empirical formula is CH_2 has a molar mass of 98.00 g/mol. What is the molecular formula of the compound?

Solution

Perform step 1:

The molar mass of the empirical formula is the sum of the masses of one carbon atom and two hydrogen atoms.

$$\text{C: } 12.011 \times 1 = 12.011$$
$$\text{H: } 1.0079 \times 2 = 2.0158$$

Molar mass of empirical formula $= 12.011 + 2.0158 = 14.027$

Perform step 2:

$$\frac{\text{Molar mass of compound}}{\text{Molar mass of empirical formula}} = \frac{98.00}{14.027} = 6.99 \approx 7$$

Perform step 3:

Multiplying the subscripts in the empirical formula by 7, we get the molecular formula, C_7H_{14}.

Example 2

A compound whose empirical formula is CH_3O has a molar mass of 62.0 g/mol. What is the molecular formula of the compound?

Solution

Perform step 1:

The molar mass of the empirical formula is the sum of the masses of one carbon atom and two hydrogen atoms.

C: $12.011 \times 1 = 12.011$

H: $1.0079 \times 3 = 3.0237$

O: $15.999 \times 1 = 15.999$

Molar mass of empirical formula = $12.011 + 3.0237 + 15.999 = 31.034$

Perform step 2:

$$\frac{\text{Molar mass of compound}}{\text{Molar mass of empirical formula}} = \frac{62.0}{31.034} = 1.998 \approx 2$$

Perform step 3:

Multiplying the subscripts in the empirical formula by 2, we get the molecular formula, $C_2H_6O_2$.

Practice Problems

8.7 A compound whose empirical formula is C_3H_3O has a molar mass of 110.0 g/mol. What is the molecular formula of the compound?

8.8 A compound whose empirical formula is CH_2 has a molar mass of 322 g/mol. What is the molecular formula of the compound?

Chart 8.5 Determining the percent composition from the molecular formula

Step 1

- Assume exactly 1 mole (unlimited number of significant digits) of compound.
- The molecular formula gives the number of moles of each element in the compound.

⬇

Step 2

- Determine the number of grams of each element in the compound by multiplying the moles of the element by its atomic mass.

- Determine the molar mass of the compound by summing the masses in grams of all the elements in the compound.

⬇

Step 3

- Divide the grams of each element by the grams of 1 mole of compound (the molar mass of the compound).
- Convert this number to a percent by multiplying by 100.

⬇

Step 4

- Check the answer by adding all of the percents from step 3 to ensure that the sum is 100% (or a number very close to 100%).

Example 1

What is the mass percent of each element in $C_3H_4O_3$?

Solution

Perform step 1:

One mole of compound contains 3 moles of C, 4 moles of H, and 3 moles of O.

Perform step 2:

Grams of C: $12.011 \times 3 = 36.033$ g C

Grams of H: $1.0079 \times 4 = 4.0316$ g H

Grams of O: $15.994 \times 3 = \underline{47.982}$ g O

Molar mass of compound = 88.047 g compound

Perform step 3:

$$\%C = \frac{36.033 \text{ g C}}{88.047 \text{ g compound}} \times 100\% = 40.92\% \text{ C}$$

$$\%H = \frac{4.0316 \text{ g H}}{88.047 \text{ g compound}} \times 100\% = 4.58\% \text{ H}$$

$$\%O = \frac{47.982 \text{ g O}}{88.047 \text{ g compound}} \times 100\% = 54.50\% \text{ O}$$

Perform step 4:

Sum of the percents = 40.92% + 4.58% + 54.50% = 100%

Example 2

What is the mass percent of each element in $Al(SO_4)_3$?

Solution

Perform step 1:

One mole of compound contains 1 mole of Al, 3 moles of S, and 12 moles of O.

Perform step 2:

Grams of Al: $26.982 \times 1 = 26.982$ g Al

Grams of S: $32.066 \times 3 = 96.198$ g S

Grams of O: $15.994 \times 12 = \underline{191.93}$ g O

Molar mass of compound = 315.11 g compound

Perform step 3:

$$\%\text{Al} = \frac{26.982 \text{ g Al}}{315.11 \text{ g compound}} \times 100\% = 8.5627\% \text{ Al}$$

$$\%\text{S} = \frac{96.198 \text{ g S}}{315.11 \text{ g compound}} \times 100\% = 30.528\% \text{ H}$$

$$\%\text{O} = \frac{191.93 \text{ g O}}{315.11 \text{ g compound}} \times 100\% = 60.909\% \text{ O}$$

Perform step 4:

Sum of the percents = $8.5627\% + 30.528\% + 60.909\% \approx 100\%$

Practice Problems

8.9 What is the mass percent of each element in SnF_4?

8.10 What is the mass percent of each element in $Mg(NO_3)_2$?

Chart 8.6 Determining the empirical formula from the percent composition

Step 1

- Assume you have exactly 100 grams (unlimited significant digits) of compound.
- In 100 grams of compound, all percents are equivalent to the number of grams of each element.

⬇

Step 2

- Convert the grams of each element to moles by dividing by the atomic mass of the element.

⬇

Step 3

- Use the calculated numbers of moles of each element as the subscripts in the empirical formula.
- Divide the values of all subscripts by the smallest one, and if the resulting values are very close to whole numbers, round them to whole numbers.
- If one or more subscripts is far from a whole number, find some multiplier that makes them whole numbers.
- Multiply all the subscripts by the multiplier that makes whole numbers.

Example 1

Determine the empirical formula of a compound that has the following percent composition: 24.39% carbon, 4.08% hydrogen, and 71.53% chlorine.

Solution

Perform step 1:

100 grams of compound contain 24.39 g carbon, 4.08 g hydrogen, and 71.53 g chlorine.

Perform step 2:

$$\text{Moles of C} = 24.39 \text{ g C} \times \frac{1 \text{ mol C}}{12.011 \text{ g C}} = 2.031 \text{ mol C}$$

$$\text{Moles of H} = 4.08 \text{ g H} \times \frac{1 \text{ mol H}}{1.0079 \text{ g H}} = 4.048 \text{ mol H}$$

$$\text{Moles of Cl} = 71.53 \text{ g Cl} \times \frac{1 \text{ mol Cl}}{35.453 \text{ g Cl}} = 2.018 \text{ mol Cl}$$

Perform step 3:

Using the moles from step 2 and dividing by the smallest number, we get $C_{\frac{2.031}{2.018}}H_{\frac{4.048}{2.018}}Cl_{\frac{2.018}{2.018}}$.

$$C_{\frac{2.031}{2.018}}H_{\frac{4.048}{2.018}}Cl_{\frac{2.018}{2.018}} \rightarrow CH_2Cl$$

The empirical formula is therefore CH_2Cl.

Example 2

Determine the empirical formula of a compound that has the following percent composition: 85.71% carbon and 14.29% hydrogen.

Solution

Perform step 1:

100 grams of compound contains 85.71 g carbon and 14.29 g hydrogen.

Perform step 2:

$$\text{Moles of C} = 85.71 \text{ g C} \times \frac{1 \text{ mol C}}{12.011 \text{ g C}} = 7.136 \text{ mol C}$$

$$\text{Moles of H} = 14.29 \text{ g H} \times \frac{1 \text{ mol H}}{1.0079 \text{ g H}} = 14.18 \text{ mol H}$$

Perform step 3:

Using the moles from step 3 and dividing by the smallest number, we get $C_{\frac{7.136}{7.136}}H_{\frac{14.18}{7.136}}$.

$$C_{\frac{7.136}{7.136}}H_{\frac{14.18}{7.136}} \rightarrow CH_2$$

The empirical formula is therefore CH_2.

Practice Problems

8.11 Determine the empirical formula of a compound that has the following percent composition: 29.71% carbon, 6.23% hydrogen, and 64.06% lead.

8.12 Determine the empirical formula of a compound that has the following percent composition: 92.30% carbon and 7.70% hydrogen.

Quiz for Chapter 8 Problems

1. In the reaction of Ca with O_2 to produce CaO, according to the unbalanced equation below, how many grams of CaO can be produced from 640.0 grams of O_2?

   $$Ca + O_2 \longrightarrow CaO$$

2. If 350.0 grams of H_2 are allowed to react with 350.0 grams of O_2, how much H_2O will be produced? The unbalanced equation for the reaction is shown below.

   $$H_2 + O_2 \longrightarrow H_2O$$

3. When 4.50 grams of a compound was subjected to combustion analysis, 10.23 grams of CO_2 and 4.18 grams of H_2O were produced. What is the empirical formula of the compound?

4. A compound is found to contain 32.13% Al and 67.87% F. Determine the empirical formula of the compound.
5. A compound has an empirical formula of CH_2O. What is the molecular formula of the compound if the molar mass is 180.0 g/mol?
6. Nitrogen gas can be prepared by a reaction that occurs according to the unbalanced equation below. How many grams of N_2 can be prepared from the reaction of 1500.0 grams of NaN_3?

   $$NaN_3 \longrightarrow Na + N_2$$

7. In the reaction of CH_4 with S to produce CS_2 and H_2S, according to the unbalanced equation below, how many grams of S are needed to exactly react with 450.0 grams of CH_4?

   $$CH_4 + S \longrightarrow CS_2 + H_2S$$

8. To determine the molecular formula of a compound, the__________ mass must be divided by the mass of the ____________ formula.
9. What is the mass percent of each element in $C_7H_{16}O$?
10. A fellow student tells you that it is easy to determine the limiting reactant in a reaction because it is the reactant that is present in the smallest number of grams. Is she correct? Explain.
11. Calculate the number of grams of $SiCl_4$ needed to exactly react with 500.0 grams of water according to the (unbalanced) equation below.

    $$SiCl_4 + H_2O \longrightarrow SiO_2 + HCl$$

12. How many grams of SiO_2 will be produced by the reaction described in problem 11?
13. Determine the empirical formula of a compound that contains C, H, and O in the following percentages: 65.45% C, 5.492% H, 29.06% O.
14. By what number must each subscript in the empirical formula be multiplied to obtain the molecular formula for a compound that has a molar mass of 582 g/mol and an empirical formula weight of 97 g/mol?
15. If 200.00 grams of Cl_2 are allowed to react with 10.00 grams of H_2 to produce HCl, which reactant is limiting? The reaction occurs by the (unbalanced) equation below:

    $$H_2 + Cl_2 \longrightarrow HCl$$

16. How many moles of HCl will be produced in problem 15? How many grams of HCl will be produced?

CHAPTER

9

The Transfer of Electrons from One Atom to Another in a Chemical Reaction

Overview: What You Should Be Able to Do

Chapter 9 provides a discussion of what happens to electrons during chemical reactions and presents a method for determining whether a reaction will occur spontaneously in the direction written or in the opposite direction. After mastering Chapter 9, you should be able to solve the following types of problems:

1. Assign oxidation states by the electron-bookkeeping method.
2. Assign oxidation states by the shortcut method.
3. Determine whether a reaction is an oxidation–reduction (redox) reaction or not.
4. Use the electromotive force series (EMF series) to write the spontaneous redox reaction that will occur between two metals and the cations of the metals.

Chart 9.1 **Assigning oxidation states by the electron-bookkeeping method**

Step 1

- Draw the correct electron dot diagram for the molecule.

⬇

Step 2

Assign every valence electron to an atom:

- Assign each lone pair to the atom on which it is drawn in the dot diagram.
- Assign all electrons in each bond to the atom with the higher electronegativity.

- If bonding electrons are shared by atoms of equal electronegativity, divide the electrons evenly between the two atoms.

⬇

Step 3

- Determine the oxidation state of each atom by subtracting the number of electrons assigned to the atom (as determined in step 2) from the number of valence electrons in a free atom of that element (as determined from its group in the periodic table).

⬇

Step 4

- Check your answer.
- The sum of the oxidation states for all the atoms should equal the charge on the molecule or ion. (The charge on a molecule is always 0 because molecules are neutral. The charge on a polyatomic ion is the charge written in the formula of the ion.)

Example 1

Use the electron-bookkeeping method to determine the oxidation state of each atom in HCl.

Solution

Perform step 1:

The correct electron dot diagram for HCl is

$$\mathrm{H}:\ddot{\underset{\cdot\cdot}{\mathrm{Cl}}}:$$

Perform step 2:

Cl will be assigned 6 electrons from its lone pairs.

Because Cl is more electronegative than hydrogen, chlorine will be assigned the 2 electrons in the H–Cl bond.

Perform step 3:

Oxidation state of each atom	=	Number of valence electrons in free atom	−	Number of electrons assigned to the atom from step 2	
H	=	1	−	0	= +1
Cl	=	7	−	8	= −1

Perform step 4:

Sum of oxidation states =

H: +1

Cl: −1

$+1 + (-1) = 0$, which is the charge on the HCl molecule.

Example 2

Use the electron-bookkeeping method to determine the oxidation state of each atom in SO_4^{2-}.

Solution

Perform step 1:

The correct electron dot diagram for SO_4^{2-} is

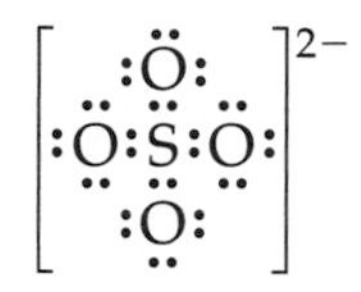

Perform step 2:

Each O will be assigned 6 electrons from the lone pairs.

Because oxygen is more electronegative than sulfur, oxygen will be assigned the 2 electrons in each S–O bond. Therefore each oxygen is assigned 8 electrons, and the sulfur assigned 0 electrons.

Perform step 3:

Oxidation state of each atom	=	Number of valence electrons in free atom	−	Number of electrons assigned to the atom from step 2	
O	=	6	−	8	= −2
S	=	6	−	0	= +6

Perform step 4:

Sum of oxidation states =
O: -2×4 (for each of four O atoms) $= -8$
S: $+6 \times 1 = +6$

Sum $= -8 + 6 = -2$, which is the charge on the SO_4^{2-} polyatomic ion.

Example 3

Use the electron-bookkeeping method to determine the oxidation state of each atom in $COCl_2$. The carbon is the central atom; the other three atoms are attached to carbon.

Solution

Perform step 1:

The correct electron dot diagram for $COCl_2$ is

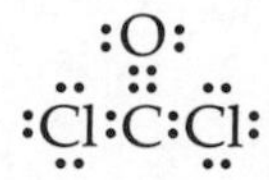

:O:
:Cl:C:Cl:

Perform step 2:

Each Cl will be assigned 6 electrons from the lone pairs.

Because chlorine is more electronegative than carbon, chlorine will be assigned the 2 electrons in each C–Cl bond. Therefore each chlorine is assigned 8 electrons.

The C is assigned 0 electrons because it has no lone pairs, and will be assigned no electrons from the bonds because carbon is less electronegative than chlorine and oxygen.

The O will be assigned 4 electrons from the lone pairs.

Because oxygen is more electronegative than carbon, oxygen will be assigned the 4 electrons in the C–O bond. Therefore oxygen is assigned 8 electrons.

Perform step 3:

Oxidation state of each atom	=	Number of valence electrons in free atom	–	Number of electrons assigned to the atom from step 2	
Cl	=	7	–	8	= −1
C	=	4	–	0	= +4
O	=	6	–	8	= −2

Perform step 4:

Sum of oxidation states =
Cl: -1×2 (for each of two Cl atoms) $= -2$
C: $+4 \times 1 = +4$
O: $-2 \times 1 = -2$

Sum $= -2 + 4 - 2 = 0$, which is the charge on the $COCl_2$ molecule.

Example 4

Use the electron-bookkeeping method to determine the oxidation state of each atom in HCN. The carbon is the central atom.

Solution

Perform step 1:

The correct electron dot diagram for HCN is

H:C:::N:

Perform step 2:

H will be assigned 0 electrons from the bonding electrons (H is less electronegative than C). C will be assigned 2 electrons from the H–C bond and 0 electrons from the C–N bond (C is less electronegative than N). Therefore the carbon is assigned a total of 2 electrons.

The N atom will be assigned 2 electrons from the lone pairs.

Because nitrogen is more electronegative than carbon, nitrogen will be assigned the 6 electrons from the C–N bond. Therefore nitrogen is assigned a total of 8 electrons.

Perform step 3:

Oxidation state of each atom	=	Number of valence electrons in free atom	−	Number of electrons assigned to the atom from step 2	
H	=	1	−	0	= +1
C	=	4	−	2	= +2
N	=	5	−	8	= −3

Perform step 4:

Sum of oxidation states =
H: +1
C: $+2 \times 1 = +2$
N: $-3 \times 1 = -3$

Sum = $+3 + (-3) = 0$, which is the charge on the HCN molecule.

Practice Problems

9.1 Use the electron-bookkeeping method to determine the oxidation state of each atom in CH_3F.

9.2 Use the electron-bookkeeping method to determine the oxidation state of each atom in H_2O.

9.3 Use the electron-bookkeeping method to determine the oxidation state of each atom in ClO_4^-.

9.4 Use the electron-bookkeeping method to determine the oxidation state of each atom in CO.

Chart 9.2 Assigning oxidation states by the shortcut method

Step 1

- Use the rules below for assigning oxidation numbers for each element in the molecule or polyatomic ion.

  1. The oxidation state of a neutral atom in its most abundant naturally occurring elemental form is typically 0.
  2. The oxidation state of an oxygen atom in a compound is almost always −2.
  3. The oxidation state of a hydrogen atom in a compound is +1.
  4. The oxidation state of any group I (1) atom in a compound is +1.
  5. The oxidation state of any group II (2) atom in a compound is +2.
  6. The oxidation state of a monatomic ion is equal to the charge on the ion.
  7. The sum of the oxidation states of the atoms in a chemical formula must add up to the overall charge on the molecule or ion represented by the formula. This rule is used to determine the oxidation number on elements for which there is no rule specified.

Halide rule:

A halogen atom in a compound usually has an oxidation state of −1.

⬇

Step 2

- Recheck your answer.
- The sum of the oxidation states for all the atoms should equal the charge on the molecule or ion. (The charge on a molecule is always 0 because molecules are neutral. The charge on a polyatomic ion is the charge written in the formula of the ion.)

Note: When assigning oxidation states using the shortcut rules, it is very useful to have two rows of numbers above each element. The bottom row of numbers will represent the oxidation state of the atom. The top row will represent the total charge contributed by the element, taking into account both the oxidation number and the number of atoms of the element. This method is presented in the examples below.

Example 1

Use the shortcut method to determine the oxidation state of each atom in HCl.

Solution

Perform step 1:

Assigning oxidation numbers by the rules, we get the following oxidation numbers:

H: + 1
Cl: − 1

Viewing the oxidation numbers and total charges in two rows, we get

Total charge: +1 −1 = 0
Oxidation number: +1 −1
HCl

Perform step 2:

Sum of oxidation states =
H: +1
Cl: −1

+1 + (−1) = 0, which is the charge on the HCl molecule.

Example 2

Use the shortcut method to determine the oxidation state of each atom in SO_4^{2-}.

Solution

Perform step 1:

Assigning oxidation numbers by the rules, we get the following oxidation numbers.

O: −2
S : must be determined by using rule 7. There is no rule for the assignment of an oxidation number to S, so the oxidation number is assigned first for O (oxidation state = −2). Because there are four O's, the total charge from the oxygens will be −8. In order for the total charge on the ion to be 2− (from the formula), the total charge from the S must be +6. Because there is only one S atom, it must have a +6 oxidation state.

Viewing the oxidation numbers and total charges in two rows, we get

Total charge: $+6 -8 = -2$
Oxidation number: $+6 \ -2$
$S \ O_4^{2-}$

Perform step 2:

Sum of oxidation states =
S: $+6 \times 1 = +6$
O: $-2 \times 4 = -8$

$+6 + (-8) = -2$, which is the charge on the SO_4^{2-} polyatomic ion.

Example 3

Use the shortcut method to determine the oxidation state of each atom in $COCl_2$.

Solution

Perform step 1:

Assigning oxidation numbers by the rules, we get the following oxidation numbers:
O: −2
Cl: −1
C: must be determined by using rule 7. There is no rule for the assignment of an oxidation number to C, so the oxidation number is assigned first for O (oxidation state = −2) and Cl (oxidation state = −1). Because there is one O, the total charge from the oxygen will be −2. Because there are two Cl's, the total charge from the chlorines will be −2. In order for the total charge on the molecule to be 0, the total charge from the C must be +4. Because there is only one C atom, it must have a +4 oxidation state.

Viewing the oxidation numbers and total charges in two rows, we get

Total charge: $+4 -2 -2 = 0$
Oxidation number: $+4 \ -2 \ -1$
$C \ O \ Cl_2$

Perform step 2:

Sum of oxidation states =
C: $+4 \times 1 = +4$
O: $-2 \times 1 = -2$
Cl: $-1 \times 2 = -2$

$+4 + (-2) + (-2) = 0$, which is the charge on the $COCl_2$ molecule.

Example 4

Use the shortcut method to determine the oxidation state of each atom in $C_4H_8O_2$.

Solution

Perform step 1:

Assigning oxidation numbers by the rules, we get the following oxidation numbers:
H: +1
O: −2
C: must be determined by using rule 7. There is no rule for the assignment of an oxidation number to C, so the oxidation number is assigned first for H (oxidation state = +1) and O (oxidation state = −2). Because there are two O's, the total charge from the oxygens will be −4. Because there are eight H's, the total charge from the hydrogens will be +8. In order for the total charge on the molecule to be 0, the total charge from the C must be −4. Because there are four C atoms, each one must have a −1 oxidation state.

Viewing the oxidation numbers and total charges in two rows, we get

Total charge:	−4	−8	−4 = 0
Oxidation number:	−1	+1	−2
	C_4	H_8	O_2

Perform step 2:

Sum of oxidation states =
C: $-1 \times 4 = -4$
H: $+1 \times 8 = +8$
O: $-2 \times 2 = -4$

$-4 + 8 + (-4) = 0$, which is the charge on the $C_4H_8O_2$ molecule.

Practice Problems

9.5 Use the shortcut method to determine the oxidation state of each atom in CH_3F.

9.6 Use the shortcut method to determine the oxidation state of each atom in H_2S.

9.7 Use the shortcut method to determine the oxidation state of each atom in ClO_4^-.

9.8 Use the shortcut method to determine the oxidation state of each atom in CO.

9.9 Use the shortcut method to determine the oxidation state of each atom in $C_2O_4^{2-}$.

Chart 9.3 Determining whether a reaction is an oxidation–reduction (redox) reaction or not

Step 1

- Assign oxidation numbers to each atom on the reactants side and each atom on the products side. (Be careful to correctly assign oxidation numbers. If numbers are incorrect, step 2 will not yield correct information.)

⬇

Step 2

- Compare oxidation numbers of each element on the reactants side with the oxidation number of the same element on the products side.
- If the oxidation number of an element on the products side is different from its oxidation number on the reactants side, a redox reaction has occurred.
- If there is no change in oxidation number for any elements, the reaction is not a redox reaction.

Example 1

Indicate whether the following reaction is a redox reaction:

$$H_2 + Cl_2 \longrightarrow 2\,HCl$$

Perform step 1:

Assigning oxidation numbers we get

Total charge:	0		0		+1 −1 = 0
Oxidation number:	0		0		+1 −1
	H_2	+	Cl_2	$\longrightarrow$	2 HCl

(Notice that the coefficients in the balanced equation have no effect on the oxidation states. You can ignore them when assigning oxidation states.)

Perform step 2:

Because there are changes in the oxidation numbers of hydrogen (0 to +1) and chlorine (0 to −1) in going from reactants to products, this reaction is a redox reaction.

Example 2

Indicate whether the following reaction is a redox reaction:

$$HCl + LiOH \longrightarrow LiCl + H_2O$$

Perform step 1:

Assigning oxidation numbers we get:

total charge:	$+1\ -1 = 0$	$+1\ -2 + 1 = 0$	$+1\ -1 = 0$	$+2\ -2 = 0$
oxidation number:	$+1\ -1$	$+1\ -2 + 1$	$+1\ -1$	$+1\ -2$
	H Cl +	Li O H $\longrightarrow$	Li Cl +	H_2 O

Perform step 2:

Because none of the atoms in this reaction have undergone a change in oxidation number in going from reactants to products, a redox reaction has not occurred.

Practice Problems

9.10 Indicate whether the following reaction is a redox reaction:

$$CaCl_2 + Na_2O \longrightarrow 2\ NaCl + CaO$$

9.11 Indicate whether the following reaction is a redox reaction:

$$2\ Al_2O_3 \longrightarrow 4\ Al + 3\ O_2$$

Chart 9.4 Using the electromotive force series (EMF series) to write the spontaneous redox reaction that will occur between two metals and the cations of the metals

Step 1

- Use the EMF series in the textbook to determine which metal is more active (higher) in the series.

⬇

Step 2

- Write the spontaneous redox reaction by writing the more active metal on the reactants side and the less active metal on the products side.
- The cation of the less active metal will be on the reactants side and the cation of the more active metal will be on the products side.

Example 1

Write the spontaneous redox reaction that will occur when a beaker of Cd^{2+} ions and Cd metal are placed in contact (through a wire) with Mg^{2+} ions and magnesium metal.

Solution

Perform step 1:

Magnesium is the more active metal, so Mg will be a reactant and Cd will be a product in the spontaneous reaction.

Perform step 2:

The spontaneous reaction will be

$$Mg + Cd^{2+} \longrightarrow Mg^{2+} + Cd$$

Example 2

Write the spontaneous redox reaction that will occur when a beaker of Ni^{2+} ions and Ni metal are placed in contact (through a wire) with Pb^{2+} ions and Pb metal.

Solution

Perform step 1:

Nickel is the more active metal, so Ni will be a reactant and Pb will be a product in the spontaneous reaction.

Perform step 2:

The spontaneous reaction will be

$$Ni + Pb^{2+} \longrightarrow Ni^{2+} + Pb$$

Practice Problems

9.12 Write the spontaneous redox reaction that will occur when a beaker of Co^{2+} ions and Co metal are placed in contact (through a wire) with Cu^{2+} ions and Cu metal.

9.13 Write the spontaneous redox reaction that will occur when a beaker of Zn^{2+} ions and Zn metal are placed in contact (through a wire) with Sn^{2+} ions and Sn metal.

Quiz for Chapter 9 Problems

1. Use the electron-bookkeeping method to assign oxidation numbers for each atom in the following species:

   (a) HI (b) CH_2Cl_2 (c) SO_3 (d) C_2H_6

2. Use the shortcut method to assign oxidation numbers for each atom in the following species:

   (a) HNO_3 (b) $S_2O_3^{2-}$ (c) $CaSO_4$

3. Indicate whether each of the following reactions is a redox reaction:
   (a) $B_2O_3 + 3\,Mg \longrightarrow 2\,B + 3\,MgO$
   (b) $HF + KOH \longrightarrow KF + H_2O$
   (c) $Zn + 2\,HCl \longrightarrow ZnCl_2 + H_2$
4. Indicate whether each of the reactions below is spontaneous in the direction written or in the opposite direction.
   (a) $Zn + Cu^{2+} \longrightarrow Cu + Zn^{2+}$
   (b) $Li + K^{+} \longrightarrow K + Li^{+}$
   (c) $Fe + Sn^{2+} \longrightarrow Sn + Fe^{2+}$
   (d) $Ni + Fe^{2+} \longrightarrow Fe + Ni^{2+}$
5. Write the spontaneous reaction that will occur when a beaker of Ni^{2+} ions and Ni metal are placed in contact (through a wire) with Mn^{2+} ions and Mn metal.
6. Complete the following sentence:
   Metals higher in the electromotive force series tend to lose electrons __________ easily than metals lower in the EMF series.
7. Indicate the oxidation number of carbon in each of the compounds below.
   (a) CO_2 (b) C_2H_6 (c) C_2H_4 (d) C_2H_2
8. Indicate the change in oxidation number for each element in the following reaction. Which element(s) undergo no change in oxidation state?

$$Mg + H_2SO_4 \longrightarrow MgSO_4 + H_2$$

9. Some mercury is accidentally dropped onto a ring that is known to be either gold or silver. If there is no reaction when the mercury is in contact with the ring, which metal is the ring made of? Explain.
10. A strip of copper metal is placed in a solution containing nickel ions. Will a reaction occur or not? Why or why not?
11. Show that the electron bookkeeping method and the shortcut method yield the same result when assigning oxidation numbers to each element in the polyatomic ion PO_4^{3-}.

Cumulative Quiz for Chapters 7, 8, and 9

1. Balance the equation for the reaction $C_6H_{12} + O_2 \longrightarrow CO_2 + H_2O$.
2. The coefficient in front of the HBr when the equation below is balanced is _____.

   $$PBr_3 + H_2O \longrightarrow H_3PO_3 + HBr$$

3. Which of the following ionic salts is (are) insoluble in water?

   (a) $BaSO_4$ (b) Na_2CO_3 (c) AgCl (d) NH_4Br

4. Write the formula of the precipitate that forms when a solution of Na_2SO_4 is mixed with a solution of PbF_2.
5. Write the molecular equation, the ionic equation, and the net ionic equation for the reaction that occurs when HBr reacts with $Ca(OH)_2$.
6. Write the molecular equation, the ionic equation, and the net ionic equation for the precipitation reaction that occurs when a solution of Na_2CO_3 is mixed with a solution of $Ba(NO_3)_2$.
7. In the reaction of Mg with N_2 to produce Mg_3N_2, how many grams of Mg_3N_2 can be produced from 85.00 grams of N_2 according to the unbalanced equation below?

   $$Mg + N_2 \longrightarrow Mg_3N_2$$

8. If 100.0 grams of H_2 are allowed to react with 250.0 grams of N_2, how much NH_3 will be produced? The unbalanced equation for the reaction is shown below.

   $$N_2 + H_2 \longrightarrow NH_3$$

9. When 1.00 grams of a compound were subjected to combustion analysis, 3.03 grams of CO_2 and 1.55 grams of H_2O were produced. What is the empirical formula of the compound?
10. A compound is found to contain 73.9% Hg and 26.1% Cl. Determine the empirical formula of the compound.
11. A compound has an empirical formula of C_3H_3O. What is the molecular formula of the compound if the molar mass is 54 g/mol?
12. What is the mass percent of each element in $KMnO_4$?

13. Calculate the number of grams of Ca needed to exactly react with 500.0 grams of water according to the following (unbalanced) equation:

$$Ca + H_2O \longrightarrow Ca(OH)_2 + H_2$$

14. If 150.00 grams of Br_2 are allowed to react with 50.00 grams of H_2 to produce HBr, which reactant is limiting? The reaction occurs by the following (unbalanced) equation:

$$H_2 + Br_2 \longrightarrow HBr$$

15. How many moles of HBr will be produced in problem 14? How many grams of HBr will be produced?
16. Use the electron bookkeeping method to assign oxidation numbers for each atom in the following species:

    (a) HBr (b) CH_3Cl (c) SO_2 (d) C_3H_8
17. Use the shortcut method to assign oxidation numbers for each atom in the following species:

    (a) $HClO_4$ (b) $Cr_2O_7^{2-}$ (c) $Mg_3(PO_4)_2$
18. Indicate whether each of the following reactions is a redox reaction:

    (a) $HBr + NaOH \longrightarrow NaBr + H_2O$

    (b) $Zn + 2\,HNO_3 \longrightarrow ZnNO_3 + H_2$
19. Indicate whether each of the reactions below is spontaneous in the direction written or in the opposite direction.

    (a) $Zn + Ca^{2+} \longrightarrow Ca + Zn^{2+}$ (b) $Na + K^{+} \longrightarrow K + Na^{+}$
20. Write the spontaneous reaction that will occur when a beaker of Zn^{2+} ions and Zn metal are placed in contact (through a wire) with Sn^{2+} ions and Sn metal.
21. Indicate the change in oxidation number for each element in the following reaction. Which element(s) undergo no change in oxidation state?

$$Zn + HCl \longrightarrow ZnCl_2 + H_2$$

22. A strip of lead is placed in a solution containing nickel ions. Will a reaction occur or not?
23. Show that the electron bookkeeping method and the shortcut method yield the same result when assigning oxidation numbers to each element in NF_3.
24. Potassium metal does not occur naturally in the Earth's crust, whereas platinum does. Explain this in terms of the relative positions of the two elements on the electromotive force (EMF) series.
25. Write the spontaneous redox reaction that will occur when a beaker of Co^{2+} ions and Sn metal are placed in contact (through a wire) with Sn^{2+} ions and Co metal.

CHAPTER

10

Intermolecular Forces and the Phases of Matter

Overview: What You Should Be Able to Do

Chapter 10 provides a discussion of the various forces that hold molecules together to form liquids and solids. After mastering Chapter 10, you should be able to solve the following types of problems:

1. Determine the type(s) of intermolecular attractive forces operating in a compound.
2. Determine the relative boiling points of substances based on their structures.

Chart 10.1 Determining the type(s) of intermolecular attractive forces operating in a compound

Step 1

- Determine whether the molecule is polar or nonpolar.
- If the compound is nonpolar, London forces are the only intermolecular attractive forces in the compound.
- If the compound is polar, proceed to step 2.

⬇

Step 2

- Determine whether the compound contains N–H, O–H, or H–F bonds.
- If there are no N–H, O–H, or H–F bonds, the compound has dipole–dipole forces in addition to London forces.

- If the compound contains N–H, O–H, or H–F bonds, the intermolecular attractive forces present are hydrogen-bonding, dipolar interactions, and London forces.
- Generally speaking, hydrogen-bonding is stronger than dipole–dipole interactions, which are stronger than London forces.

Example 1

What type(s) of intermolecular attractive forces is (are) present in dimethyl ether, CH_3OCH_3?

Solution

Perform step 1:

The molecule is polar.

Perform step 2:

There are no N–H, O–H, or H–F bonds. Therefore the compound has dipole–dipole and London forces present.

Example 2

What type(s) of intermolecular attractive forces is (are) present in CBr_4?

Solution

Perform step 1:

The molecule is nonpolar.

Perform step 2:

Therefore London forces are the only forces present in the compound.

Example 3

What type(s) of intermolecular attractive forces is (are) present in methylamine, CH_3NH_2?

Solution

Perform step 1:

The molecule is polar.

Perform step 2:

There are N–H bonds. Therefore the compound has hydrogen bonds, dipolar forces, and London forces.

Practice Problems

10.1 What type(s) of intermolecular attractive forces is (are) present in CH_2F_2?

10.2 What type(s) of intermolecular attractive forces is (are) present in CH_3OH?

Chart 10.2 Determining the relative boiling points of substances based on their structures

Step 1

- Determine the types of intermolecular attractive forces present in the compounds being ranked.
- If the compounds have different intermolecular attractive forces present, use the general rule that hydrogen-bonding is stronger than dipole–dipole forces, which are stronger than London forces.
- The stronger the intermolecular attractive forces, the higher the boiling point.
- If all compounds have the same type(s) of intermolecular forces, proceed to step 2.

⬇

Step 2

- Consider the relative sizes of the molecules of the compounds being compared.
- In general, the larger the molecule the higher the boiling point if the compounds all have the same type(s) of intermolecular forces.

Example 1

Arrange the following compounds in order of increasing boiling points:

CH_3OCH_3, $CH_3CH_2CH_3$, CH_3CH_2OH

Solution

Perform step 1:

CH_3OCH_3 has dipole–dipole and London intermolecular attractive forces.

$CH_3CH_2CH_3$ has only London forces.

CH_3CH_2OH has hydrogen-bonding, dipole–dipole forces, and London forces.

Perform step 2.

Therefore $CH_3CH_2CH_3$ has the lowest boiling point and CH_3CH_2OH has the highest. The arrangement of compounds in order of increasing boiling points is:

$$CH_3CH_2CH_3 < CH_3OCH_3 < CH_3CH_2OH$$

Example 2

Arrange the following compounds in order of increasing boiling points:

$CH_3CH_2CH_3$ $\quad$ CH_3CH_3 $\quad$ CH_4

Solution

Perform step 1:

All three compounds are nonpolar and have only London forces.

Perform step 2:

The larger the molecule, the more electrons and the stronger the intermolecular attractive forces. The order is therefore,

$$CH_4 < CH_3CH_3 < CH_3CH_2CH_3$$

Practice Problems

10.3 Which has a higher boiling point, CH_3Cl or CH_3OH?

10.4 Arrange the following compounds in order of increasing boiling points:

$CH_3CH_2CH_2CH_3$ $\quad$ $CH_3CH_2CH_2NH_2$ $\quad$ $CH_3CH_2OCH_3$

Quiz for Chapter 10 Problems

1. Indicate the type(s) of intermolecular attractive forces in CH_3Cl.
2. Two compounds, A and B, have boiling points of 78°C and 30°C, respectively. If the two compounds are known to be CH_3OCH_3 and CH_3CH_2OH, which compound has which boiling point? Explain.
3. Which compound in each pair has the lower boiling point?
   (a) CH_4 and CH_3Cl $\quad$ (b) CH_3CH_3OH and CH_3CH_2Cl
4. Indicate the type(s) of intermolecular attractive forces operating in each type of compound below.
   (a) nonpolar
   (b) polar but without N–H, O–H, and H–F bonds
   (c) polar and with N–H, O–H, or H–F bonds
5. For compounds that have the same type(s) of intermolecular attractive forces present, the larger the molecule, the ____________ the boiling point.
6. Which compound in each pair has stronger intermolecular attractive forces?
   (a) CH_3CH_3 and $CH_3CH_2CH_2CH_3$
   (b) CH_3OH and CH_3CH_2OH

7. Which molecule(s) below can participate in hydrogen bonding?

$CH_3OCH_2CH_3$ $\quad$ $CH_3N(CH_3)CH_3$ $\quad$ CH_3NH_2 $\quad$ H_2S

8. Which molecules have dipole–dipole intermolecular attractive forces?

(a) CF_4 (b) CH_2Cl_2 (c) CH_3OH (d) BF_3

9. Arrange the following compounds in order of increasing boiling points:

(a) CH_3CH_3 (b) CH_3CH_2Cl (c) CH_3OH (d) CH_3F (e) CH_4

10. Which compound has the higher boiling point, CH_3OH or CH_3NH_2? Explain.

CHAPTER

11

What If There Were No Intermolecular Forces? The Ideal Gas

Overview: What You Should Be Able to Do

Chapter 11 provides a description of the gaseous state and presents the mathematical description of a gas in the form of the ideal gas equation. After mastering Chapter 11, you should be able to solve the following types of problems:

1. Use the ideal gas equation to find the value of P, V, T, or n when the values of the other three are given.
2. Solve for the molar mass of a gas using the molar mass form of the ideal gas equation.
3. Solve initial-condition, final-condition problems.
4. Solve gas stoichiometry problems using the ideal gas law and a balanced chemical equation.

Chart 11.1 Using the ideal gas equation to find the value of P, V, T, or n when the values of the other three are given

Step 1

- Algebraically solve the ideal gas equation for the value to be determined.

⬇

Step 2

- Make sure that the given values are in the correct units for use in the ideal gas equation (P in atmospheres, n in moles, v in liters, T in kelvins). Convert any values that are not in the correct units.

⬇

Step 3

- Substitute the given values into the equation obtained in step 1 and perform the arithmetic operation to arrive at the solution.

Example 1

A gas sample contains 3.20 moles of gas at a temperature of 30.0°C and a pressure of 832 mm Hg. What volume does the gas occupy?

Solution

Perform step 1:

$$V = \frac{nRT}{P}$$

Perform step 2:

P must be converted to atm and T must be converted to K.

$$\text{atm} = 832 \text{ mm Hg} \times \frac{1 \text{ atm}}{760 \text{ mm Hg}} = 1.09 \text{ atm}$$

$$\text{K} = 30.0°\text{C} + 273.15 = 303.2 \text{ K}$$

Perform step 3:

$$V = \frac{(3.20 \text{ mol})(0.0821 \text{ L}\cdot\text{atm}/\text{mole}\cdot\text{K})(303.2 \text{ K})}{1.09 \text{ atm}} = 73.1 \text{ L}$$

Example 2

A container of compressed oxygen gas has a volume of 12.0 L. The pressure of the compressed gas is 53.7 atm, and the temperature of the gas is 28.4°C. How many moles of gas are present?

Solution

Perform step 1:

$$n = \frac{PV}{RT}$$

Perform step 2:

T must be converted to K.

$$K = 28.4°C + 273.15 = 301.6 \text{ K}$$

Perform step 3:

$$n = \frac{(53.7 \text{ atm})(12.0 \text{ L})}{(0.0821 \text{ L}\cdot\text{atm} / \text{mol}\cdot\text{K})(301.6 \text{ K})} = 26.0 \text{ moles of gas}$$

Practice Problems

11.1 What is the pressure of 0.823 moles of N_2 gas in a 5.00-L container at 37.0°C?

11.2 What is the temperature of 4.0 moles of a gas in a 3.50-L steel container at a pressure of 98.2 atm?

11.3 Calculate the moles of gas in a 12.00-g sample of gas that occupies 7.4 L at 27°C and 980 mm Hg.

Chart 11.2 Solving for the molar mass of a gas using the molar mass (*MM*) form of the ideal gas equation

Step 1

- Write the ideal gas equation in the molar mass form,

$$MM = \frac{m_{\text{sample}} RT}{PV}$$

⬇

Step 2

- Make sure that the given values are in the correct units for use in the ideal gas equation (P in atmospheres, n in moles, v in liters, T in kelvins). Convert any values that are not in the correct units.

⬇

Step 3

- Substitute the given values into the equation and perform the arithmetic operation to arrive at the molar mass.

Example 1

A 24.5-g sample of an unknown gas occupies a volume of 44.5 L at 273.0°C and 0.100 atm. What is the molar mass of the sample?

Solution

Perform step 1:

$$MM = \frac{m_{\text{sample}}RT}{PV}$$

Perform step 2:

T must be converted to kelvins.

$$\text{K} = 273°\text{C} + 273.15 = 546.2 \text{ K}$$

Perform step 3:

$$MM = \frac{(24.5 \text{ g})(0.0821 \text{ L} \cdot \text{atm} / \text{mol} \cdot \text{K})(546.2 \text{ K})}{(0.100 \text{ atm})(44.5 \text{ L})} = 247 \text{ g} / \text{mol}$$

Example 2

What is the molar mass of an unknown gas if 12.04 g of the gas occupies 7.40 L at 27.0°C and 980 mm Hg?

Solution

Perform step 1:

$$MM = \frac{m_{\text{sample}}RT}{PV}$$

Perform step 2:

T must be converted from °C to K.

$$\text{K} = 27.0°\text{C} + 273.15 = 300.2 \text{ K}$$

P must be converted from mm Hg to atm.

$$P = 980 \text{ mm Hg} \times \frac{1 \text{ atm}}{760 \text{ mm Hg}} = 1.29 \text{ atm}$$

Perform step 3:

$$MM = \frac{(12.04 \text{ g})(0.0821 \text{ L} \cdot \text{atm} / \text{mol} \cdot \text{K})(300.2 \text{ K})}{(1.29 \text{ atm})(7.40 \text{ L})} = 31.1 \text{ g} / \text{mol}$$

Practice Problems

11.4 Calculate the molar mass of a gas if 1.00 L of the gas has a mass of 5.38 g at 15°C and 736 mm Hg.

11.5 What is the molar mass of a gas if 0.985 g of the gas occupies 3.00 L at a pressure of 178 mm Hg and a temperature of 22.5°C?

Chart 11.3 Solving initial-condition, final-condition problems

Step 1

- List P, V, n, and T for the initial and final conditions.
- Write in numeric values for any known variables.
- Where possible, express the final variable in terms of the initial one (e.g., $V_f = V_i$).

⬇

Step 2

- Write the expression

$$\frac{P_iV_i}{n_iT_i} = \frac{P_fV_f}{n_fT_f}$$

- Whenever possible, rewrite this expression to show final variables in terms of initial ones (from step 1).
- Cancel factors that are identical on the two sides of the expression.

⬇

Step 3

- Solve the equation algebraically for the desired variable and substitute the numerical values for the known quantities.
- Perform the calculation to get the answer.

Example 1

7.62 moles of gas in a container with a moveable piston are heated from 100.0 °C to 175.0°C. If the pressure remains constant, what volume will the gas occupy after the heating if it occupied 2.50 L initially?

Solution

Perform step 1:

Initial conditions

$P_i = P_f$ (unchanged)
$V_i = 2.50$ L
$n_i = n_f = 7.62$ mol (unchanged)
$T_i = 373.2$ K

Final conditions

$P_f = P_i$ (unchanged)
$V_f = ?$
$n_f = n_i = 7.62$ mol (unchanged)
$T_f = 448.2$ K

Perform step 2:

$$\frac{\cancel{P_i}V_i}{\cancel{n_i}T_i} = \frac{\cancel{P_f}V_f}{\cancel{n_f}T_f} \longrightarrow \frac{V_i}{T_i} = \frac{V_f}{T_f}$$

Perform step 3:

$$\frac{V_i}{T_i} = \frac{V_f}{T_f} \longrightarrow V_f = \frac{T_f V_i}{T_i}$$

$$V_f = \frac{(448.2 \text{ K})(2.50 \text{ L})}{373.2 \text{ K}} = 3.00 \text{ L}$$

This answer agrees with the ideal gas rules. When the temperature increases, the volume will increase if the pressure and the number of moles are held constant.

Example 2

What happens to the pressure inside a steel container if the temperature of the container is decreased from 25.3°C to 5.2°C, and the number of moles of gas inside the container is increased from 2.32 mol to 8.63 mol? The volume of the container is 40.0 L and the initial pressure within the container is 170.0 atm.

Solution

Perform step 1:

Initial conditions

$P_i = 170.0$ atm
$V_i = V_f = 40.0$ L (unchanged)
$n_i = 2.32$ mol
$T_i = 298.5$ K

Final conditions

$P_f = ?$
$V_f = V_i = 40.0$ L (unchanged)
$n_f = 8.63$ mol
$T_f = 278.4$ K

Perform step 2:

$$\frac{P_i V_i}{n_i T_i} = \frac{P_f V_f}{n_f T_f} \longrightarrow \frac{P_i}{n_i T_i} = \frac{P_f}{n_f T_f}$$

Perform step 3:

$$\frac{P_i}{n_i T_i} = \frac{P_f}{n_f T_f} \longrightarrow P_f = \frac{P_i n_f T_i}{n_i T_i}$$

$$P_f = \frac{(170.0 \text{ atm})(8.63 \text{ mol})(278.4 \text{ K})}{(2.32 \text{ mol})(298.5 \text{ K})} = 590 \text{ atm}$$

Practice Problems

11.6 A quantity of gas occupying 2.10 L at a pressure of 750 mm Hg is compressed to a volume of 1.20 L. Calculate the resulting pressure of the gas, assuming that the temperature and the number of moles are held constant.

11.7 A sample of gas occupying a volume of 8.20 L at a temperature of 20.0°C and a pressure of 0.832 atm is heated to 50.0°C and its pressure is increased to 9.47 atm. Calculate the final volume of the gas.

11.8 If 3.00 moles of a gas occupy 65.3 L at 25°C and 1.05 atm pressure, what volume will the gas occupy at 10°C and a pressure of 0.83 atm?

Chart 11.4 Solving gas stoichiometry problems using the ideal gas law and a balanced chemical equation

Step 1

- Use the ideal gas equation to determine the number of moles of gas.

⬇

Step 2

- Use the coefficients in the balanced equation from the reaction to determine the number of moles of any other substance in the equation.

⬇

Step 3

- Use the molar mass of the substance if asked to determine the number of grams of the other substance.

Example 1

Pure oxygen can be produced by heating mercury(II) oxide, HgO. The balanced equation is given below.

$$2\,HgO(s) \longrightarrow 2\,Hg(l) + O_2(g)$$

How many moles of HgO will be needed to produce 45.0 L of oxygen gas at a temperature of 22.4°C and a pressure of 1.23 atm?

Solution

Perform step 1:

Use the ideal gas equation to find the moles of oxygen. (Remember to convert °C to K.)

$$n = \frac{PV}{RT}$$

$$n = \frac{(1.23\ \text{atm})(45.0\ \text{L})}{(0.0821\ \text{L}\cdot\text{atm}/\text{mol}\cdot\text{K})(295.6\ \text{K})} = 2.28\ \text{mol}\ O_2$$

Perform step 2:

$$\text{Moles HgO} = 2.28 \text{ mol } O_2 \times \frac{2 \text{ mol HgO}}{1 \text{ mol } O_2} = 4.6 \text{ mol HgO}$$

Example 2

How many moles of HCl will be needed to produce 300.0 L of H_2 gas at a temperature of 25°C and a pressure of 1.00 atm? The balanced equation for the reaction is shown below.

$$Zn + 2\ HCl \longrightarrow H_2 + ZnCl_2$$

Solution

Perform step 1:

Use the ideal gas equation to find the moles of hydrogen. (Remember to convert °C to K.)

$$n = \frac{PV}{RT}$$

$$n = \frac{(1.00 \text{ atm})(300.0 \text{ L})}{(0.0821 \text{ L} \cdot \text{atm} / \text{mol} \cdot \text{K})(298.2 \text{ K})} = 12.3 \text{ mol } H_2$$

Perform step 2:

$$\text{Moles HCl} = 12.3 \text{ mol } H_2 \times \frac{2 \text{ mol HCl}}{1 \text{ mol } H_2} = 24.6 \text{ mol HCl}$$

Example 3

How many grams of C_8H_{18} will be needed to produce 250.0 L of CO_2 gas at a temperature of 0.00°C and a pressure of 752 mm Hg? The balanced equation for the reaction is shown below.

$$2\ C_8H_{18} + 25\ O_2 \longrightarrow 16\ CO_2 + 18\ H_2O$$

Solution

Perform step 1:

Use the ideal gas equation to find the moles of CO_2. (Remember to convert °C to K and mm Hg to atm.)

$$n = \frac{PV}{RT}$$

$$n = \frac{(0.989 \text{ atm})(250.0 \text{ L})}{(0.0821 \text{ L} \cdot \text{atm} / \text{mol} \cdot \text{K})(273.15 \text{ K})} = 11.0 \text{ mol } CO_2$$

Perform step 2:

$$\text{Moles } C_8H_{18} = 11.0 \text{ mol } CO_2 \times \frac{2 \text{ mol } C_8H_{18}}{16 \text{ mol } CO_2} = 1.40 \text{ mol } C_8H_{18}$$

Perform step 3:

$$\text{Grams of } C_8H_{18} = 1.40 \text{ mol } C_8H_{18} \times \frac{114.2 \text{ g } C_8H_{18}}{1 \text{ mol } C_8H_{18}} = 1.60 \times 10^2 \text{ g } C_8H_{18}$$

Practice Problems

11.9 How many moles of P_4 are needed to produce 50.0 L of PH_3 gas at 28.0°C and 2.0 atm pressure? The balanced equation for the process is shown below.

$$P_4 + 6\,H_2 \longrightarrow 4\,PH_3$$

11.10 How many grams of CO_2 will be produced from the reaction of 5175.0 L of C_3H_8 with oxygen gas at 35.0°C and 1.73 atm pressure? The balanced equation for the process is shown below.

$$C_3H_8 + 5\,O_2 \longrightarrow 3\,CO_2 + 4\,H_2O$$

11.11 How many moles of O_2 will be needed to produce 179.0 L of CO_2 gas at 25.0°C and 1.03 atm pressure? The balanced equation for the process is shown below.

$$C_3H_8 + 5\,O_2 \longrightarrow 3\,CO_2 + 4\,H_2O$$

Quiz for Chapter 11 Problems

1. What volume will 3.64 moles of N_2 gas occupy at 25.0°C and 4.32 atm?
2. How many moles of gas are present in a 5.0-L container that is being held at a temperature of 10.0°C and a pressure of 109.0 atm?
3. Calcium carbonate can be heated to produce carbon dioxide gas according to the following reaction. How many grams of $CaCO_3$ are required for the production of 1000.0 L of CO_2 at 37°C and 1.06 atm?

$$CaCO_3 \longrightarrow CaO + CO_2$$

4. Fill in the blanks below.
   (a) When the pressure is doubled on a sample of gas occupying 6.80 L, the new volume (holding the temperature constant) will be ______.
   (b) When the temperature (absolute) is doubled on a sample of gas occupying 6.80 L, the new volume (holding the pressure constant) will be ______.
5. What volume will 8.92 grams of Ne gas occupy at 22.0°C and 782 mm Hg?
6. At what temperature (in °C) will a 4.00-mol sample of gas exert a pressure of 39.0 atm in a 2.50-L container?
7. If a gas occupies a volume of 4.21 L at 50°C, what volume will the gas occupy at 100°C? Assume the number of moles and the pressure are held constant.
8. What pressure will be exerted by 5.00 moles of a gas initially at 32.0°C and 8.2 atm if the temperature of this quantity of gas is decreased to 20.0°C?

9. Oxygen gas can be produced by the decomposition of hydrogen peroxide according to the following reaction:

$$H_2O_2 \longrightarrow H_2O + O_2$$

(a) Balance the equation for the reaction.

(b) How many moles of H_2O_2 are required to produce 15.0 L of O_2 gas at a temperature of 295 K and a pressure of 740.0 mm Hg?

10. Which sample contains more moles of gas?

(a) 5.00 L of He at 22.0°C and 2.0 atm pressure

(b) 2.50 grams of He

11. How many grams of KCl are needed to produce 780.0 L of O_2 at 24.0°C and 1.20 atm pressure according to the following reaction? (*Hint:* Make sure the equation is balanced.)

$$KClO_3 \longrightarrow KCl + O_2$$

12. Calculate the molar mass of a gas if 2.68 grams of the gas occupies 2.00 L at 0°C and 764 mm Hg.

CHAPTER

12

Solutions

Overview: What You Should Be Able to Do

Chapter 12 provides a discussion of solutions, including solution formation and concentration units. After mastering Chapter 12, you should be able to solve the following types of problems:

1. Prepare a solution from solute and solvent (from scratch).
2. Prepare a solution from a more concentrated stock solution.
3. Calculate the percent composition of a solution.
4. Solve solution stoichiometry problems.
5. Solve acid–base titration problems.
6. Determine molar mass using boiling-point elevation (ΔT_b) or freezing-point depression (ΔT_f) data.

Chart 12.1 Preparing a solution from solute and solvent (from scratch)

Step 1

- Determine the moles of solute needed to prepare the solution.
- Moles of solute = Volume of solution (in liters) × Molarity of solution

⬇

Step 2

- Convert the moles of solute to grams of solute.
- Use the molar mass as a conversion factor.

⬇

Step 3

- Place the number of grams from step 2 in a volumetric flask and add enough water to bring the total volume of the solution to the desired volume.
- Do not simply add an amount of water equal to the desired volume of solution.

Example 1

Explain how you would prepare 2.50 L of a 0.350 M solution of glucose, $C_6H_{12}O_6$.

Solution

Perform step 1:

$$\begin{aligned}\text{Moles of glucose needed} &= \text{Liters of solution} \times \text{Molarity of solution}\\ &= 2.50\ \text{L} \times 0.350\ \text{mol glucose/L}\\ &= 0.875\ \text{mol glucose}\end{aligned}$$

Perform step 2:

$$\text{Grams of glucose} = 0.875\ \text{mol glucose} \times \frac{180.15\ \text{g glucose}}{1\ \text{mol glucose}} = 158\ \text{g glucose}$$

Perform step 3:

Place 158 grams of glucose in a 2.50-L volumetric flask, and add enough water to bring the level of the solution to 2.50 L.

Example 2

Explain how you would prepare 500.0 mL of a 1.20 M solution of potassium nitrate, KNO_3.

Solution

Perform step 1:

$$\begin{aligned}\text{Moles of } KNO_3 \text{ needed} &= \text{Liters of solution} \times \text{Molarity of solution}\\ &= 0.500\ \text{L} \times 1.20\ \text{mol } KNO_3\text{/L}\\ &= 0.600\ \text{mol } KNO_3\end{aligned}$$

Perform step 2:

$$\text{Grams of glucose} = 0.600\ \text{mol } KNO_3 \times \frac{101.1\ \text{g } KNO_3}{1\ \text{mol } KNO_3} = 60.6\ \text{g } KNO_3$$

Perform step 3:

Place 60.6 grams of KNO_3 in a 500.0-mL volumetric flask and add enough water to bring the level of the solution to 500.0 mL.

Practice Problems

12.1 Explain how you would prepare 1.00 L of a 4.50 M solution of NaCl.

12.2 Explain how you would prepare 250.0 mL of a 0.250 M solution of sucrose, $C_{12}H_{22}O_{11}$.

Chart 12.2 Method A: Preparing a solution from a more concentrated stock solution

Step 1

- Determine the moles of solute needed to prepare the solution.
- Moles of solute = Volume of solution (in liters) × Molarity of solution

⬇

Step 2

- Convert the moles of solute to volume of stock solution by using the molarity of the stock solution as a conversion factor. (Multiply the moles of solute by the inverse of the stock solution's molarity.)

⬇

Step 3

- Place the volume of stock solution from step 2 in a volumetric flask and add enough water to bring the total volume of the solution to the desired volume.

Method B: Preparing a solution from a more concentrated stock solution

Step 1

- Use the dilution equation shown below to determine the volume of the stock solution needed:

$$\text{Molarity}_{\text{stock solution}} \times \text{Volume}_{\text{stock solution}} = \text{Molarity}_{\text{diluted solution}} \times \text{Volume}_{\text{diluted solution}}$$

- Algebraically solve the equation for the volume of the stock solution.

⬇

Step 2

- Place the volume of stock solution from step 2 in a volumetric flask and add enough water to bring the total volume of the solution to the desired volume.

Example 1

Explain how you would prepare 500.0 mL of a 0.100 M solution of glucose from a stock solution that is 0.250 M.

Solution (by method A)

Perform step 1:

$$\begin{aligned}\text{Moles of glucose needed} &= \text{Volume of solution (L)} \times \text{Molarity of solution}\\ &= 0.500\ \text{L} \times 0.100\ \text{mol/L}\\ &= 0.0500\ \text{mol glucose}\end{aligned}$$

Perform step 2:

$$\text{Volume of stock solution} = \text{Moles of solute needed} \times \frac{1}{\text{Molarity of stock solution}}$$

$$= 0.0500\ \text{mol} \times 1/0.250\ \text{mol/L}$$

$$= 0.200\ \text{L} = 200\ \text{mL stock solution}$$

Perform step 3:

Place 200 mL of stock solution in a 500-mL volumetric flask and bring the solution level up to the 500 mL mark.

Solution (by method B)

Perform step 1:

$$\text{Molarity}_{\text{stock solution}} \times \text{Volume}_{\text{stock solution}} = \text{Molarity}_{\text{diluted solution}} \times \text{Volume}_{\text{diluted solution}}$$

$$(0.250\ \text{mol/L})(V_{\text{stock solution}}) = (0.500\ \text{mol/L})(0.100\ \text{L})$$

$$V_{\text{stock solution}} = \frac{(0.500\ \text{mol/L})(0.100\ \text{L})}{0.250\ \text{mol/L}} = 0.200\ \text{L} = 2.00 \times 10^2\ \text{mL}$$

Perform step 2:

Place 2×10^2 (200) mL of stock solution in a 500.0-mL volumetric flask and bring the solution level up to the desired volume.

Example 2

Explain how you would prepare 2.00 L of a 0.750 M HCl solution from a stock solution that is 6.0 M.

Solution (by method A)

Perform step 1:

$$\begin{aligned}\text{Moles of HCl needed} &= \text{Volume of solution (L)} \times \text{Molarity of solution}\\ &= 2.00\ \text{L} \times 0.750\ \text{mol/L}\\ &= 1.50\ \text{mol HCl}\end{aligned}$$

Perform step 2:

$$\text{Volume of stock solution} = \text{Moles of solute needed} \times \frac{1}{\text{Molarity of stock solution}}$$

$$= 1.50 \text{ mol} \times 1/6.00 \text{ mol/L}$$

$$= 0.250 \text{ L} = 250 \text{ mL stock solution}$$

Perform step 3:

Place 250 mL of stock solution in a 2.00-L volumetric flask and bring the solution level up to the desired volume.

Solution (by method B)

Perform step 1:

$$\text{Molarity}_{\text{stock solution}} \times \text{Volume}_{\text{stock solution}} = \text{Molarity}_{\text{diluted solution}} \times \text{Volume}_{\text{diluted solution}}$$

$$(6.00 \text{ mol/L})(V_{\text{stock solution}}) = (0.750 \text{ mol/L})(2.00 \text{ L})$$

$$V_{\text{stock solution}} = \frac{(0.750 \text{ mol/L})(2.00 \text{ L})}{6.00 \text{ mol/L}}$$

$$= 0.250 \text{ L} = 250 \text{ mL}$$

Perform step 2:

Place 250 mL of stock solution in a 2.00-L volumetric flask and bring the solution level up to the desired volume.

Practice Problems

12.3 Explain how you would prepare 1.00 L of a 2.50 M NaCl solution from a stock solution that is 4.50 M.

12.4 Explain how you would prepare 250.0 mL of a 12.00 M H_2SO_4 solution from a stock solution that is 18.00 M.

Chart 12.3 Calculating the percent composition of a solution

Step 1

- Determine the type of percent composition calculation the problem involves (i.e., percent by mass, percent by volume, or percent by mass/volume).

⬇

Step 2

- Calculate the percent composition using the appropriate formula.
- $\text{Percent by mass (mass \%)} = \frac{\text{Grams of solute}}{\text{Grams of solution}} \times 100$
- $\text{Percent by volume (vol \%)} = \frac{\text{Volume of solute}}{\text{Volume of solution}} \times 100$
- $\text{Percent by mass / volume (mass / vol \%)} = \frac{\text{Grams of solute}}{\text{Volume of solution}} \times 100$

Example 1

Calculate the percent by mass of a solution that contains 125.0 grams of $KClO_3$ in 450.0 grams of solution.

Solution

Perform step 1:

The problem asks for percent by mass, so we will use

$$\text{Percent by mass (mass \%)} = \frac{\text{Grams of solute}}{\text{Grams of solution}} \times 100$$

Perform step 2:

$$\% \ KClO_3 \text{ by mass} = \frac{125.0 \text{ g } KClO_3}{450.0 \text{ g solution}} \times 100 = 27.78\% \ KClO_3$$

Example 2

Calculate the percent by mass of a solution that contains 55.0 grams of NaBr in 150.0 grams of water.

Solution

Perform step 1:

The problem asks for percent by mass, so we will use

$$\text{Percent by mass (mass \%)} = \frac{\text{Grams of solute}}{\text{Grams of solution}} \times 100$$

Note that we must add the mass of the solute, NaBr, and the mass of the solvent, water, to get the total mass of the solution. Therefore, the mass of the solution equals 55.0 grams (NaBr) + 150.0 grams (water), or 205.0 grams of solution.

Perform step 2:

$$\% \text{ NaBr by mass} = \frac{55.0 \text{ g } KClO_3}{205.0 \text{ g solution}} \times 100 = 26.8\% \text{ NaBr}$$

Example 3

Calculate the percent by mass/volume of a solution that contains 80.0 grams of $LiNO_3$ in 550.0 mL of solution.

Solution

Perform step 1:

The problem asks for percent by mass/volume, so we will use

$$\text{Percent by mass / volume (mass / vol \%)} = \frac{\text{Grams of solute}}{\text{Milliliters of solution}} \times 100$$

Perform step 2:

$$\% \; LiNO_3 \text{ by mass / vol} = \frac{80.0 \text{ g } LiNO_3}{550.0 \text{ mL solution}} \times 100 = 14.5\% \; LiNO_3$$

Practice Problems

12.5 Calculate the percent by volume of a solution that contains 15.0 mL of methanol, CH_3OH, in 120.0 mL of solution.

12.6 Calculate the percent by mass/volume of a solution that contains 65.0 grams of sucrose, $C_{12}H_{22}O_{11}$, in 550.0 mL of solution.

Chart 12.4 Solving solution stoichiometry problems

Step 1

- Write the balanced equation for the reaction that is occurring in the solution.

⬇

Step 2

- Convert all given amounts to moles using molarities (if volumes are given) and/or molar masses (if grams are given) as conversion factors.

⬇

Step 3

- Use the coefficients from the balanced equation to find the relationship between the number of moles of the given substance and the number of moles of the unknown substance.

- The ratio of the coefficients is used as a conversion factor, with the coefficient in front of the unknown as the numerator and the coefficient in front of the given as the denominator.
- The conversion factor is multiplied by the number of moles of the given substance.

⬇

Step 4

- Convert the moles of the unknown, calculated in step 3, to appropriate units using molarities (if volumes are asked for) and/or molar masses (if grams are asked for) as conversion factors.

Example 1

How many milliliters of 0.300 M NaOH are required to react exactly with 50.0 mL of 0.200 M H_2SO_4 according to the reaction represented by the unbalanced equation below?

$$H_2SO_4 + NaOH \longrightarrow Na_2SO_4 + H_2O$$

Solution

Perform step 1:

$$H_2SO_4 + 2\,NaOH \longrightarrow Na_2SO_4 + 2\,H_2O$$

Perform step 2:

$$\text{Moles of } H_2SO_4 = \text{Molarity} \times \text{Volume (L)} = 0.200 \text{ mol } H_2SO_4\,/\text{L } H_2SO_4 \times 0.050 \text{ L } H_2SO_4 = 0.0100 \text{ mol } H_2SO_4$$

Perform step 3:

$$\text{Moles of NaOH} = 0.0100 \text{ mol } H_2SO_4 \times \frac{2 \text{ mol NaOH}}{1 \text{ mol } H_2SO_4} = 0.0200 \text{ mol NaOH}$$

Perform step 4:

$$\text{Liters of NaOH} = \frac{0.0200 \text{ mol NaOH}}{0.300 \text{ mol NaOH / L}} = 0.0667 \text{ L NaOH}$$

Volume of NaOH in milliliters = 66.7 mL NaOH

Example 2

What is the maximum number of grams of $MgCl_2$ that can be prepared from reacting 350.0 mL of 0.125 M NaCl with an excess amount of $Mg(NO_3)_2$ according to the unbalanced equation below for the reaction?

$$NaCl + Mg(NO_3)_2 \longrightarrow MgCl_2 + NaNO_3$$

Solution

Perform step 1:

$$2\,NaCl + Mg(NO_3)_2 \longrightarrow MgCl_2 + 2\,NaNO_3$$

Perform step 2:

$$\text{Moles of NaCl} = \text{Molarity} \times \text{Volume (L)} = 0.125\ \text{mol NaCl/L NaCl} \times 0.350\ \text{L NaCl} = 0.0438\ \text{mol NaCl}$$

Perform step 3:

$$\text{Moles of } MgCl_2 = 0.0438\ \text{mol NaCl} \times \frac{1\ \text{mol } MgCl_2}{2\ \text{mol NaCl}} = 0.0219\ \text{mol } MgCl_2$$

Perform step 4:

$$\text{Grams of } MgCl_2 = 0.0219\ \text{mol } MgCl_2 \times \frac{95.211\ \text{g } MgCl_2}{1\ \text{mol } MgCl_2} = 2.09\ \text{g } MgCl_2$$

Example 3

How would you prepare 25.00 g of $CaCO_3$ from a 0.400 M solution of Ca_3N_2 and a 0.100 M solution of Na_2CO_3? The unbalanced reaction is given below.

$$Ca_3N_2 + Na_2CO_3 \longrightarrow CaCO_3 + Na_3N$$

Solution

Perform step 1:

$$Ca_3N_2 + 3\,Na_2CO_3 \longrightarrow 3\,CaCO_3 + 2\,Na_3N$$

Perform step 2:

$$\text{Moles of } CaCO_3 = 25.00\ \text{g } CaCO_3 \times \frac{1\ \text{mol}}{100.09\ \text{g}} = 0.2498\ \text{mol } CaCO_3$$

Perform step 3:

$$\text{Moles of } Ca_3N_2 = 0.2498\ \text{mol } CaCO_3 \times \frac{1\ \text{mol } Ca_3N_2}{3\ \text{mol } CaCO_3} = 0.08327\ \text{mol } Ca_3N_2$$

$$\text{Moles of } Na_2CO_3 = 0.2498\ \text{mol } CaCO_3 \times \frac{3\ \text{mol } Na_2CO_3}{3\ \text{mol } CaCO_3} = 0.2498\ \text{mol } Na_2CO_3$$

Perform step 4:

$$\text{Volume (L) of } Ca_3N_2 = \frac{\text{Moles of } Ca_3N_2}{\text{Molarity}} = \frac{0.08327\ \text{mol } Ca_3N_2}{0.400\ \text{mol / L}} = 0.208\ \text{L} = 208\ \text{mL}$$

$$\text{Volume (L) of } Na_2CO_3 = \frac{\text{Moles of } Na_2CO_3}{\text{Molarity}} = \frac{0.2498\ \text{mol } Na_2CO_3}{0.100\ \text{mol / L}} = 2.50\ \text{L} = 2500\ \text{mL}$$

To prepare 25.00 g of $CaCO_3$, combine 208 mL of 0.400 M Ca_3N_2 and 2500 mL of 0.100 M Na_2CO_3.

Practice Problems

12.7 How many grams of SiH_4 can be prepared by reacting 50.0 mL of 1.25 M HCl with an excess of Mg_2Si according to the reaction represented by the unbalanced equation below?

$$Mg_2Si + HCl \longrightarrow MgCl_2 + SiH_4$$

12.8 How would you prepare 45.0 grams of PbC_2O_4 from a 0.250 M solution of $Pb(NO_3)_2$ and a 0.125 M solution of $K_2C_2O_4$? The reaction is represented by the unbalanced equation below.

$$Pb(NO_3)_2 + K_2C_2O_4 \longrightarrow KNO_3 + PbC_2O_4$$

Chart 12.5 Solving acid–base titration problems

Step 1

- Write the balanced equation for the reaction.

⬇

Step 2

- Use the molarity to convert the volume of the base used to reach the equivalence point to moles.
- Volume (L) × Molarity = Moles

⬇

Step 3

- Use the coefficients in the balanced equation to convert moles of base added to moles of acid that reacted.
- The number of moles of acid that reacted is the number of moles of acid originally in the flask.

⬇

Step 4

- Convert the moles of the acid, calculated in step 3, to the appropriate units.
- If the molarity of the acid is asked for, divide the moles of acid by the volume of the original acid solution (before the titration began) because molarity = moles/liters.

Example 1

50.0 mL of an H_2SO_4 solution of unknown concentration is titrated with 0.100 M NaOH. 32.30 mL of base is required to neutralize the acid. What is the concentration of the H_2SO_4?

Solution

Perform step 1:

$$H_2SO_4 + 2\,NaOH \longrightarrow Na_2SO_4 + 2\,H_2O$$

Perform step 2:

$$\begin{aligned}\text{Moles of NaOH} &= \text{Volume of NaOH} \times \text{Molarity of NaOH}\\ &= 0.032\,30\ \text{L} \times 0.100\ \text{mol NaOH/L}\\ &= 0.003\,230\ \text{mol NaOH}\end{aligned}$$

Perform step 3:

$$\text{Moles of } H_2SO_4 = 0.003\,230\ \text{mol NaOH} \times \frac{1\ \text{mol } H_2SO_4}{2\ \text{mol NaOH}} = 0.001\,615\ \text{mol } H_2SO_4$$

Perform step 4:

$$\text{Molarity of } H_2SO_4 = \frac{0.001\,615\ \text{mol } H_2SO_4}{0.050\ \text{L NaOH}} = 0.0323\ \text{M } H_2SO_4$$

Example 2

50.0 mL of an HCl solution of unknown concentration is titrated with 0.200 M NaOH. 28.92 mL of base is required to neutralize the acid. What is the concentration of the HCl?

Solution

Perform step 1:

$$HCl + NaOH \longrightarrow NaCl + H_2O$$

Perform step 2:

$$\begin{aligned}\text{Moles of NaOH} &= \text{Volume of NaOH} \times \text{Molarity of NaOH}\\ &= 0.02892\ \text{L} \times 0.200\ \text{mol NaOH/L}\\ &= 0.005784\ \text{mol NaOH}\end{aligned}$$

Perform step 3:

$$\text{Moles of HCl} = 0.005\,784\ \text{mol NaOH} \times \frac{1\ \text{mol HCl}}{1\ \text{mol NaOH}} = 0.005\,784\ \text{mol HCl}$$

Perform step 4:

$$\text{Molarity of HCl} = \frac{0.005\,784 \text{ mol HCl}}{0.0500 \text{ L}} = 0.116 \text{ M HCl}$$

Practice Problems

12.9 50.0 mL of an HCl solution of unknown concentration is titrated with 0.100 M NaOH. 32.33 mL of base is required to neutralize the acid. What is the concentration of the HCl?

12.10 50.0 mL of an H_2SO_4 solution of unknown concentration is titrated with 0.250 M NaOH. 39.36 mL of base is required to neutralize the acid. What is the concentration of the H_2SO_4?

Chart 12.6 Determining molar mass using boiling-point elevation (ΔT_b) or freezing-point depression (ΔT_f) data

Step 1

- Use the boiling-point elevation equation (or the freezing-point depression equation), shown below in terms of the boiling-point elevation, to determine the moles of solute.

$$\Delta T_b = K_b \times \frac{\text{Moles of solute}}{\text{Kilograms of solvent}}$$

$$\text{Moles of solute} = \frac{\Delta T_b \times \text{Kilograms of solvent}}{K_b}$$

- For water, $K_b = 0.52°\text{C} \cdot \text{kg/mol}$ and $K_f = 1.86°\text{C} \cdot \text{kg/mol}$.
- The value of K_b or K_f will be given in the problem if the solvent is not water.
- Remember to convert the mass of solvent to kilograms if it is given in grams.

⬇

Step 2

- Divide the grams of solute (given in the problem) by the moles of solute (calculated in step 1) to determine the molar mass.

Example 1

A solution of an unknown compound was prepared by dissolving 1.50 grams of the compound in 30.0 grams of water. The freezing point of the solution was determined to be −1.02°C. What is the molar mass of the unknown compound?

Solution

Perform step 1:

$$\text{Moles of solute} = \frac{\Delta T_f \times \text{Kilograms of solvent}}{K_f}$$

$$\text{Moles of solute} = \frac{1.02\ \text{C}^\circ \times 0.030\ \text{kg H}_2\text{O}}{1.86\ \text{C}^\circ \cdot \text{kg H}_2\text{O} / \text{mol solute}} = 0.0165\ \text{mol solute}$$

Perform step 2:

$$\text{Molar mass of solute} = \frac{1.50\ \text{g compound}}{0.0165\ \text{mol compound}} = 90.1\ \text{g} / \text{mol}$$

Example 2

44.00 grams of an unknown compound was dissolved in 100.0 grams of water. The boiling point of the solution was determined to be 101.27°C. What is the molar mass of the unknown compound?

Solution

Perform step 1:

$$\text{Moles of solute} = \frac{\Delta T_b \times \text{Kilograms of solvent}}{K_b}$$

$$\text{Moles of solute} = \frac{1.27\ \text{C}^\circ \times 0.100\ \text{kg H}_2\text{O}}{0.52\ \text{C}^\circ \cdot \text{kg H}_2\text{O} / \text{mol solute}} = 0.244\ \text{mol solute}$$

Perform step 2:

$$\text{Molar mass of solute} = \frac{44.00\ \text{g compound}}{0.244\ \text{mol compound}} = 180\ \text{g} / \text{mol}$$

Practice Problem

12.11 2.20 grams of an unknown compound was dissolved in 10.0 g of water. The freezing point of the solution was determined to be −2.33°C. What is the molar mass of the unknown compound?

Quiz for Chapter 12 Problems

1. How many grams of $BaCl_2$ can be produced from the reaction of 82.5 mL of a 0.175 M NaCl solution with an excess of $Ba(OH)_2$? The reaction is represented by the unbalanced equation below.

$$\text{NaCl} + \text{Ba(OH)}_2 \longrightarrow \text{BaCl}_2 + \text{NaOH}$$

2. Describe how you would prepare 100.0 mL of a 0.420 M $C_2H_2O_6$ solution from a stock solution whose concentration is 5.00 M.
3. Calculate the volume percent of a solution that contains 35.0 mL of ethanol in 750.0 mL of solution.
4. How many milliliters of a 0.123 M HCl solution are required to react exactly with 35.32 mL of a 0.282 M NaOH solution?
5. 50.00 mL of H_2SO_4 solution is titrated to neutralization with 22.34 mL of 0.120 M NaOH. Calculate the concentration of the H_2SO_4 solution.
6. Calculate the mass percent of a solution that contains 80.0 grams of KNO_3 and 250.0 grams of water.
7. When 10.00 grams of an unknown compound are dissolved in 250.0 grams of water, the freezing point of the solution is −3.70°C. What is the molar mass of the compound?
8. Calculate the mass percent of a solution that contains 400.0 grams of KBr in 1.000 kg of solution.
9. Calculate the mass/volume percent in a solution that contains 8.50 grams of C_3H_6O in 125.0 mL of solution.
10. Describe how you would prepare 500.0 mL of a 0.631 M Na_2SO_4 solution from solid Na_2SO_4 and water.

Cumulative Quiz for Chapters 10, 11, and 12

1. For molecules that contain an H–F, N–H, or O–H bond, the strongest intermolecular attractive forces present are ________________ ____________.
2. The phase change in which a liquid changes to a gas is called ______________.
3. The melting points of nonpolar solids are generally ________than those of polar solids.
4. Indicate which substance(s) below would *not* involve hydrogen-bonding.

   (a) $CH_3OCH_2CH_3$ (b) CH_3NH (c) HF (d) CH_3OH
5. Indicate all of the intermolecular attractive forces operating in the following substances.

   (a) HF (b) CHF_3 (c) He
6. Explain why BF_3 is a nonpolar molecule, even though it has polar bonds.
7. What volume will 8.92 moles of O_2 gas occupy at 27.0°C and 2.68 atm pressure?
8. Calcium carbonate can be heated to produce carbon dioxide gas according to the following reaction. How many grams of CaO are produced from the reaction of 780.0 grams of $CaCO_3$? How many milliliters of CO_2 will be produced at 37°C and 1.06 atm pressure?

   $$CaCO_3 \longrightarrow CaO + CO_2$$
9. When the pressure is tripled on a sample of gas occupying 3.42 L, the new volume (holding the temperature constant) will be _________.
10. What volume will 425.0 grams of Ar gas occupy at 24.0°C and 779 mm Hg?
11. At what temperature (in K) will a sample of 10.0 moles of gas exert a pressure of 11.0 atm in a 45.0-L container?
12. If a gas occupies a volume of 8.50 L at 50°C, what volume will the gas occupy at 100°C? Assume the number of moles and the pressure are held constant.
13. Which sample contains more moles of gas, 5.00 L of He at 22.0°C and 2.0 atm pressure or 250.0 grams of He?
14. The solute in a solution containing 70.0 mL of methanol and 45.0 mL of water is ________________.

15. Which of the following compounds is (are) insoluble in nonpolar solvents?

    (a) BCl_3 (b) CH_4 (c) SO_2 (d) CO_2

16. Calculate the percent by mass of solute in a solution made by mixing 25.0 grams of solute and 100.0 grams of solvent.

17. Calculate the number of milliliter of 0.016 M H_2CO_3 needed to react with 50.0 milliliters of 0.121 M NaOH to produce Na_2CO_3. (*Hint:* Write the balanced equation for the reaction.)

18. When 25.0 g of an unknown substance is dissolved in 75.0 g of water, the freezing point of the solution is −5.00°C. What is the molar mass of the unknown?

19. How many moles of HNO_3 are needed to neutralize 75.0 milliliters of 0.100 M $Ca(OH)_2$?

20. Explain why CH_3CH_2OH has a higher boiling point than CH_3OCH_3 even though they have the same molar mass.

CHAPTER

13

When Reactants Turn into Products

Overview: What You Should Be Able to Do

Chapter 13 provides a discussion of the factors that influence the rate at which chemical reactions will occur. After mastering Chapter 13, you should be able to solve the following types of problems:

1. Draw a reaction-energy profile and calculate ΔE_{rxn} when given the energy of the reactants and products and the activation energy E_a of a reaction.
2. Write the rate law for a chemical reaction and determine the order of the reaction when given experimental data.
3. Determine a possible mechanism of a reaction when provided with the overall reaction and the rate law.

Chart 13.1 Drawing a reaction-energy profile and calculating ΔE_{rxn} when given the energy of the reactants and products and the activation energy E_a of a reaction

Step 1

- Draw the x- and y-axis of a graph and label the x-axis "Reaction Coordinate" and the y-axis "Energy."

⬇

Step 2

- Place the energy values on the y-axis. Place the energy of the reactants on the left portion of the x-axis, and place the energy of the products on the right portion of the x-axis.

⬇

Step 3

- Place the energy of the transition state between the energy of the reactants and the energy of the products.

⬇

Step 4

- Draw a curve connecting the energy of the reactants to that of the transition state, and the energy of the transition state to that of the products.
- Determine the ΔE_{rxn} (if it is not provided in the problem) by subtracting the energy of the reactants from the energy of the products of the reaction ($E_{products} - E_{reactants}$).
- The reaction is endothermic if the energy of the products is greater than the energy of the reactants, and the reaction is exothermic if the energy of the products is less than the energy of the reactants.

Example 1

In a hypothetical chemical reaction, the energy of the reactants is 200 kJ/mol, the energy of the products is 250 kJ/mol, and the energy of the transition state is 400 kJ/mol. Draw the reaction-energy profile for the reaction. Is the reaction endothermic or exothermic?

Solution

Perform step 1:

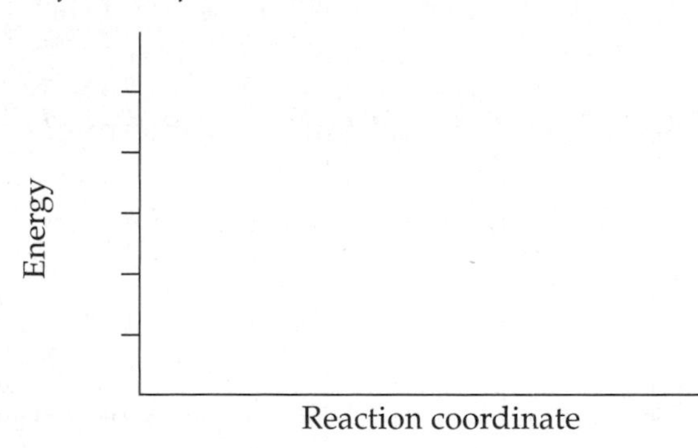

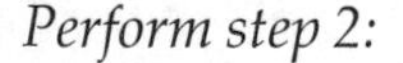

Perform step 2:

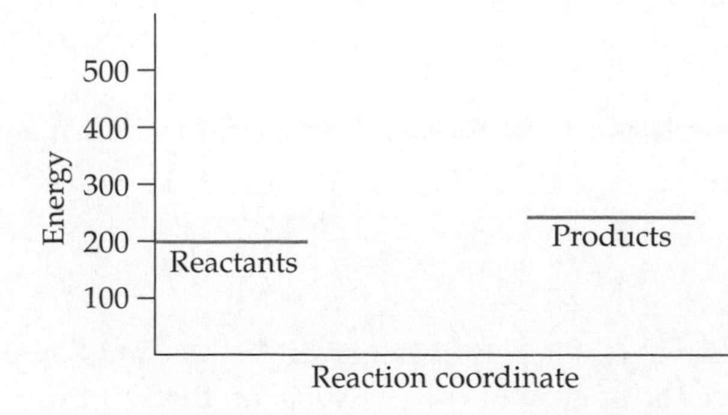

Perform step 3:

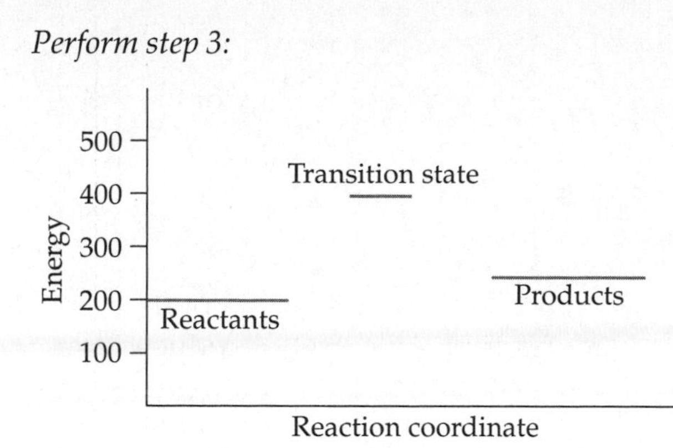

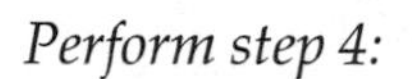

Perform step 4:

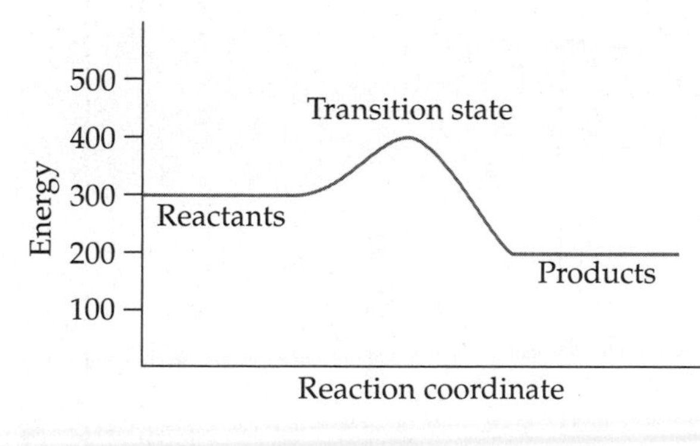

The reaction is endothermic.

Example 2

In a hypothetical chemical reaction, the energy of the reactants is 200 kJ/mol, the energy of the products is 150 kJ/mol, and the energy of the transition state is 250 kJ/mol. Draw the reaction energy profile for the reaction. Is the reaction endothermic or exothermic?

Solution

Perform step 1:

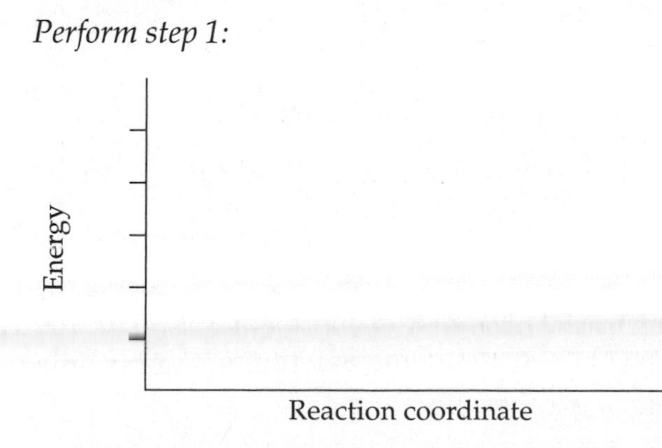

Perform step 2:

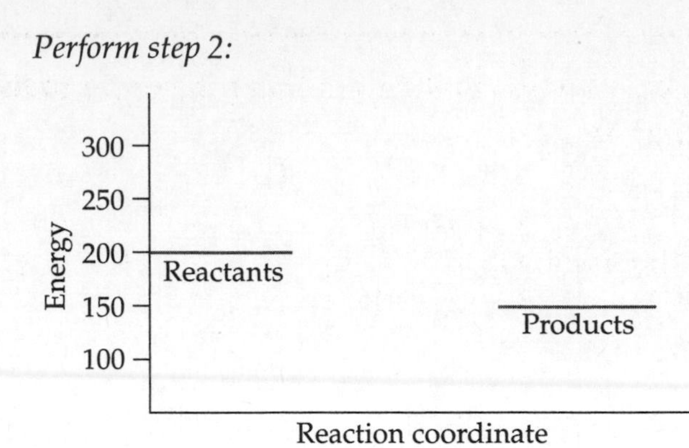

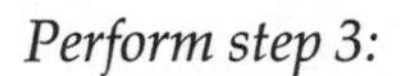
Perform step 3:

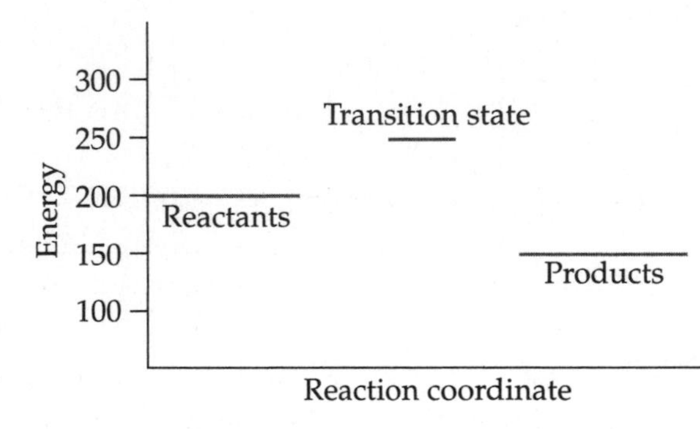

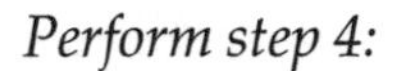
Perform step 4:

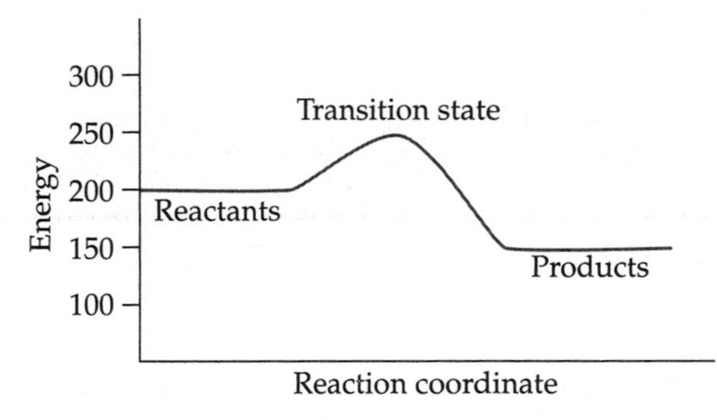

The reaction is exothermic.

Practice Problems

13.1 In the reaction X ⟶ Y, the energy of X is 50 kJ/mol, the energy of Y is 30 kJ/mol, and the energy of the transition state is 80 kJ/mol. Draw the reaction-energy profile for the reaction. Is the reaction endothermic or exothermic?

13.2 In a certain chemical reaction, the energy of the reactants is 35 kJ/mol, the energy of the products is 45 kJ/mol, and the energy of the transition state is 50 kJ/mol. Draw the reaction-energy profile for the reaction. Is the reaction endothermic or exothermic?

Chart 13.2 Writing the rate law for a chemical reaction and determining the order of the reaction when given experimental data

Step 1

- Write the general rate law for the reaction.
- The general rate law for a reaction $A + B + C \longrightarrow D$ is

$$\text{Rate} = k[A]^x[B]^y[C]^z,$$

where x, y, and z are determined from the experimental data.

⬇

Step 2

- Determine the values of x, y, and z as indicated below.
- Determine the value of x by identifying two experiments for which [A] changes but [B] and [C] remain the same.
- Determine how the rate changes with the change in [A] to get the value of x.
- Determine the value of y by identifying two experiments for which [B] changes but [A] and [C] remain the same. Determine how the rate changes with the change in [B] to get the value of y.
- Determine the value of z by identifying two experiments for which [C] changes but [A] and [B] remain the same. Determine how the rate changes with the change in [C] to get the value of z.

⬇

Step 3

- Determine the order of the reaction by adding the values of x, y, and z.

Example 1

In a kinetics study of the reaction $A + B + C \longrightarrow D$, the following data were obtained. Write the rate law for the reaction, and determine the overall order of the reaction.

Experiment	[A]	[B]	[C]	Rate (M/s) of D formation
1	0.100	0.100	0.100	0.032
2	0.100	0.200	0.100	0.128
3	0.200	0.100	0.100	0.064
4	0.100	0.100	0.200	0.064

Solution

Perform step 1:

General rate law:

$$\text{Rate} = k[A]^x[B]^y[C]^z$$

Perform step 2:

(a) Use experiments 1 and 3 to find the value of x. When [A] doubles, holding [B] and [C] constant, the rate also doubles. Because $2^1 = 2$, the value of $x = 1$.

(b) Use experiments 1 and 2 to find the value of y. When [B] doubles, holding [A] and [C] constant, the rate increases by a factor of 4. Because $2^2 = 4$, the value of $y = 2$.

(c) Use experiments 1 and 4 to find the value of z. When [C] doubles, holding [A] and [B] constant, the rate increases by a factor of 2. Because $2^1 = 2$, the value of $z = 1$.

The rate law is therefore

$$\text{Rate} = k[\text{A}]^1[\text{B}]^2[\text{C}]^1$$

Perform step 3:

The sum of x, y, and $z = 1 + 2 + 1 = 4$. Therefore the overall order of the reaction is 4.

Example 2

The rate data for the reaction $\text{E} + \text{F} \longrightarrow \text{G}$ are shown below. Write the rate law for the reaction and determine the overall order of the reaction.

Experiment	[E]	[F]	Rate (M/s) of G formation
1	0.010	0.015	0.00138
2	0.010	0.045	0.00414
3	0.030	0.015	0.01242

Solution

Perform Step 1:

General rate law:

$$\text{Rate} = k[\text{E}]^x[\text{F}]^y$$

Perform step 2:

(a) Use experiments 1 and 3 to find the value of x. When [E] is tripled (0.030/0.010 = 3), holding [F] constant, the rate increases by a factor of 9 (0.012 42/0.001 38 = 9). Because $3^2 = 9$, the value of $x = 2$.

(b) Use experiments 1 and 2 to find the value of y. When [F] triples, holding [E] constant, the rate triples (0.004 14/0.001 38 = 3). Because $3^1 = 3$, the value of $y = 1$.

Therefore

$$\text{Rate} = k[\text{E}]^2[\text{F}]^1$$

Perform step 3:

The sum of x and $y = 1 + 2 = 3$. Therefore the overall order of the reaction is 3.

Practice Problem

13.3 The results of kinetics experiments for the reaction of $R + S \longrightarrow Z$ are shown below. Write the rate law for the reaction and indicate the overall order of the reaction.

Experiment	[R]	[S]	Rate (M/s) of Z formation
1	0.240	0.240	0.060
2	0.720	0.240	0.180
3	0.240	0.480	0.120

Chart 13.3 Determining a possible mechanism of a reaction when provided with the overall reaction and the rate law

Step 1

- Write the slow step in the mechanism when provided with the rate law and the overall reaction.
- The species whose concentrations are represented in the rate law are the reactants in the slow step of the mechanism.
- The exponents on the concentrations are the coefficients in the slow step.
- The products of the slow step should reflect the same number and kind of atoms as the reactants in the slow step, but arranged to make different substances.

⬇

Step 2

- Add additional steps to the mechanism such that the sum of all steps equals the overall reaction.
- No elementary steps should have more than two reactants.

Example 1

Determine a possible mechanism for the reaction $A_2 + B \longrightarrow A_2B$. The rate law for the reaction is

$$\text{Rate} = k[A_2][Z]$$

Solution

Perform step 1:

The reactants in the slow step of the reaction are A_2 and Z because they appear in the rate law. The coefficients for both species are 1 because the exponent of each is 1 in the rate law. A possible first step in the mechanism is therefore

$$A_2 + Z \longrightarrow AZ + A$$

Perform step 2:

Adding additional steps such that the sum of all steps equals the overall reaction, a possible mechanism is

$$\begin{array}{rll} \textbf{Step 1:} & A_2 + \cancel{Z} \longrightarrow \cancel{AZ} + \cancel{A} & \text{(slow)} \\ \textbf{Step 2:} & \cancel{A} + B \longrightarrow \cancel{AB} & \text{(fast)} \\ \textbf{Step 3:} & \cancel{AZ} + \cancel{AB} \longrightarrow \cancel{Z} + A_2B & \text{(fast)} \\ \hline \text{Overall reaction:} & A_2 + B \longrightarrow A_2B & \end{array}$$

Practice Problem

13.4 Determine a possible mechanism for the reaction

$$Br_2 + CHBr_3 \longrightarrow HBr + CBr_4$$

The rate law for the reaction is

$$\text{Rate} = k[Br_2].$$

Quiz for Chapter 13 Problems

1. Fill in the blanks in the statement below.

   When the energy of the _________ is greater than the energy of the ___________ the reaction is exothermic, and when the energy of the _________ is greater than the energy of the ___________ the reaction is endothermic.

2. The reaction producing CO_2 and NO from CO and NO_2 is shown below.

   $$CO + NO_2 \longrightarrow CO_2 + NO$$

   The reaction is first order with respect to both CO and NO_2.

   (a) Write the rate law for the reaction.

   (b) What is the overall order of the reaction?

   (c) What will happen to the reaction rate when the concentration of CO is doubled?

3. Draw a reaction-energy profile diagram for a hypothetical chemical reaction with the following characteristics. The energy of the reactants is 20 kJ/mol, the energy of the products is 30 kJ/mol, and the energy of the transition state is 40 kJ/mol. Is the reaction endothermic or exothermic? Explain.

4. Write the general rate law for the following reaction, using x and y as orders:

   $$2\ NO + O_2 \longrightarrow 2\ NO_2$$

5. The rate law for the reaction $2\ NO + Br_2 \longrightarrow 2\ NOBr$ is

$$\text{Rate} = k[NO]^2[Br_2]$$

How will the rate of the reaction change when the following changes occur?

(a) [NO] is doubled (b) $[Br_2]$ is tripled

(c) [NO] is tripled (d) [NO] doubles and $[Br_2]$ triples

6. A reaction releases 700 kJ of energy.

(a) Is the reaction endothermic or exothermic?

(b) In the reaction-energy profile, are the energies of the reactants or products higher? Explain.

7. Determine a possible mechanism for the reaction

$$2\ NO + Cl_2 \longrightarrow 2\ NOCl$$

The rate law for the reaction is

$$\text{Rate} = k[NO][Cl_2]$$

8. The rate law for the reaction of $3\ A + B \longrightarrow D$ is Rate $= k[A][B]^2$. What is the overall order of the reaction?

9. The slow step in the mechanism of the reaction of H_2 with ICl to produce HCl and I_2 is shown below. Write the rate law for the reaction.

$$H_2 + ICl \longrightarrow HI + HCl$$

10. Write the rate law for the reaction $A + 2\ B \longrightarrow D$ if the following is true:

The rate of the reaction quadruples when the concentration of A is doubled, and the rate of the reaction triples when the concentration of B is tripled.

CHAPTER

14

Chemical Equilibrium

Overview: What You Should Be Able to Do

Chapter 14 presents an introduction to the state of chemical equilibrium, a state in which the rate of the forward reaction equals the rate of the reverse reaction. After mastering Chapter 14, you should be able to solve the following types of problems:

1. Determine the value of the equilibrium constant K_{eq} by using the equilibrium concentrations.
2. Calculate the K_{sp} from the solubility.
3. Calculate the solubility from the K_{sp} value. (*Note:* the solubility is also sometimes referred to as the saturation solubility or the maximum solubility.)
4. Determine the missing equilibrium concentration from the K_{eq} value and the values of all other equilibrium concentrations.

Chart 14.1 Determining the value of the equilibrium constant K_{eq} by using the equilibrium concentrations

Step 1

- Write the equilibrium constant expression for the reaction.

⬇

Step 2

- Substitute the molar equilibrium concentrations for all species in the reaction.

⬇

Step 3

- Perform the calculation to determine the value of K_{eq}.

Example 1

The reaction involving the production of ammonia gas is

$$N_2 + 3\,H_2 \rightleftarrows 2\,NH_3$$

At a certain temperature, when the reaction is at equilibrium the concentrations of the species are the following:

$$[N_2] = 0.005\,61\text{ M};\ [H_2] = 0.813\text{ M};\ [NH_3] = 0.241\text{ M}$$

Calculate the value of K_{eq}.

Solution

Perform step 1:

$$K_{eq} = \frac{[NH_3]^2}{[N_2][H_2]^3}$$

Perform step 2:

$$K_{eq} = \frac{[NH_3]^2}{[N_2][H_2]^3} = \frac{(0.241)^2}{(0.00561)(0.813)^3}$$

Perform step 3:

$$K_{eq} = \frac{(0.241)^2}{(0.00561)(0.813)^3} = 19.3$$

Example 2

For the reaction $H_2 + I_2 \rightleftarrows 2\,HI$, the equilibrium concentrations are as follows:

$$[H_2] = 0.022\text{ M};\ [I_2] = 0.36\text{ M};\ [HI] = 0.62\text{ M}$$

Calculate the value of K_{eq} for the reaction.

Solution

Perform step 1:

$$K_{eq} = \frac{[HI]^2}{[H_2][I_2]}$$

Perform step 2:

$$K_{eq} = \frac{[HI]^2}{[H_2][I_2]} = \frac{(0.62)^2}{(0.022)(0.36)}$$

Perform step 3:

$$K_{eq} = \frac{(0.62)^2}{(0.022)(0.36)} = 49$$

Practice Problems

14.1 For the hypothetical reaction $2\,A + B \rightleftarrows 2\,C$, calculate the value of K_{eq} if the following concentrations are present at equilibrium:

$[A] = 0.090$ M; $[B] = 0.072$ M; $[C] = 0.011$ M

14.2 For the reaction $2\,SO_2 + O_2 \rightleftarrows 2\,SO_3$, the following concentrations are present at equilibrium:

$[SO_2] = 0.014$ M; $[O_2] = 0.031$ M; $[SO_3] = 0.24$ M

Calculate the value of K_{eq} for the reaction.

Chart 14.2 Calculating the K_{sp} from the solubility

Step 1

- Write the equation for the equilibrium reaction of the salt dissolving in water.
- Write the equilibrium-constant K_{sp} expression from the equation.

⬇

Step 2

- Substitute the molar concentrations of the ions in solution into the K_{sp} expression.
- Use the coefficients in the solubility reaction to determine the concentration of species whose concentrations are not given in the problem.

⬇

Step 3

- Perform the calculation to determine the value of K_{sp}.

Example 1

Sparingly soluble $PbCl_2$ dissolves in water to yield a $[Pb^{2+}]$ of 0.039 M. Calculate the value of K_{sp} for the reaction.

Solution

Perform step 1:

$PbCl_2(s) \rightleftarrows Pb^{2+}(aq) + 2\,Cl^-(aq)$

$K_{sp} = [Pb^{2+}][Cl^-]^2$

Perform step 2:

$[Pb^{2+}] = 0.039$ M

$[Cl^-] = 2\,[Pb^{2+}] = 0.078$ M

$K_{sp} = [Pb^{2+}][Cl^-]^2 = (0.039)(0.078)^2$

Perform step 3:

$K_{sp} = (0.039)(0.078)^2 = 2.4 \times 10^{-4}$

Example 2

Sparingly soluble CuI dissolves in water to yield a $[Cu^+]$ of 2.26×10^{-6} M. Calculate the value of K_{sp} for the reaction.

Solution

Perform step 1:

$CuI(s) \rightleftarrows Cu^+(aq) + I^-(aq)$

$K_{sp} = [Cu^+][I^-]$

Perform step 2:

$[Cu^+] = 2.26 \times 10^{-6}$ M

$[I^-] = [Cu^+] = 2.26 \times 10^{-6}$ M

$K_{sp} = [Cu^+][I^-] = (2.26 \times 10^{-6})(2.26 \times 10^{-6})$

Perform step 3:

$K_{sp} = (2.26 \times 10^{-6})(2.26 \times 10^{-6}) = 5.1 \times 10^{-12}$

Practice Problems

14.3 When the sparingly soluble salt BaF_2 dissolves in water, the $[Ba^{2+}] = 7.5 \times 10^{-3}$ M. Calculate the K_{sp} for BaF_2.

14.4 When sparingly soluble $BaCO_3$ dissolves in water (to produce Ba^{2+} and $CO_3{}^{2-}$ ions) the $[Ba^{2+}] = 8.9 \times 10^{-5}$. Calculate the K_{sp} for $BaCO_3$.

Chart 14.3 Calculating the solubility from the K_{sp} value

Step 1

- Write the equation for the equilibrium reaction of the salt dissolving in water.
- Write the equilibrium constant K_{sp} expression from the equation.

⬇

Step 2

- Set the solubility of the compound equal to the variable s.
- Express the equilibrium concentrations of the ions in terms of s, using the coefficients in the balanced equation.
- Table 14.2 in the textbook provides the ion solubilities in terms of s for various salts.

⬇

Step 3

- Substitute the equilibrium ion concentrations, expressed in terms of s, into the K_{sp} expression.
- Solve for the value of s.

Example 1

What is the maximum solubility of $CdCO_3$ in water at 25°C? The K_{sp} for $CdCO_3$ at 25°C is 1.8×10^{-14}.

Solution

Perform step 1:

$CdCO_3(s) \rightleftarrows Cd^{2+}(aq) + CO_3^{2-}(aq)$
$K_{sp} = [Cd^{2+}][CO_3^{2-}]$

Perform step 2:

s = solubility of $CdCO_3$

Because 1 mole of $CdCO_3$ produces 1 mole each of Cd^{2+} and CO_3^{2-}, $[Cd^{2+}] = [CO_3^{2-}] = s$.

Perform step 3:

$K_{sp} = [Cd^{2+}][CO_3^{2-}]$
$1.8 \times 10^{-14} = s \times s = s^2$

Taking the square root of both sides to solve for s, we obtain $s = 1.3 \times 10^{-7}$. Therefore the maximum solubility of $CdCO_3$ in water at 25°C is 1.3×10^{-7} M.

Example 2

What is the maximum solubility of CuBr in water at 25°C? The K_{sp} for CuBr at 25°C is 4.2×10^{-8}.

Solution

Perform step 1:

$CuBr(s) \rightleftarrows Cu^{+}(aq) + Br^{-}(aq)$

$K_{sp} = [Cu^{+}][Br^{-}]$

Perform step 2:

s = solubility of CuBr

Because 1 mole of CuBr produces 1 mole each of Cu^{+} and Br^{-}, $[Cu^{+}] = [Br^{-}] = s$.

Perform step 3:

$K_{sp} = [Cu^{+}][Br^{-}]$

$4.2 \times 10^{-8} = s \times s = s^2$

Taking the square root of both sides to solve for s, we obtain $s = 2.0 \times 10^{-4}$. Therefore the maximum solubility of CuBr in water at 25°C is 2.0×10^{-4} M.

Practice Problems

14.5 Calculate the maximum solubility for AgBr at 25°C. The K_{sp} for AgBr at 25°C is 7.7×10^{-13}.

Chart 14.4 Determining the missing equilibrium concentration from the K_{eq} value and the values of all other equilibrium concentrations

Step 1

- Write the equilibrium constant K_{eq} expression for the reaction.

⬇

Step 2

- Substitute the value of K_{eq} and the molar concentrations for all known species into the K_{sp} expression.

⬇

Step 3

- Algebraically solve the equation to determine the value of the missing equilibrium concentration.

Example 1

For the reaction $2\,A_2 + B_2 \rightleftarrows 2\,A_2B$, the $K_{eq} = 3.36 \times 10^{18}$ and the equilibrium concentrations of A_2 and A_2B are as follows:

$$[A_2] = 0.0224 \text{ M}; \quad [A_2B] = 0.596 \text{ M}.$$

Calculate the $[B_2]$.

Solution

Perform step 1:

$$K_{eq} = \frac{[A_2B]^2}{[A_2]^2[B_2]}$$

Perform step 2:

$$K_{eq} = \frac{[A_2B]^2}{[A_2]^2[B_2]}$$

$$3.36 \times 10^{18} = \frac{(0.596)^2}{(0.0224)^2[B_2]}$$

Perform step 3:

$$3.36 \times 10^{18} = \frac{(0.596)^2}{(0.0224)^2[B_2]}$$

Multiplying both sides of the equation by $[B_2]$, and dividing both sides of the equation by 3.36×10^{18}, we obtain $[B_2] = 2.11 \times 10^{-16}$ M.

Practice Problems

14.6 For the reaction $N_2 + O_2 \rightleftarrows 2\,NO$, the $K_{eq} = 4.03 \times 10^{-2}$ and the equilibrium concentrations of N_2 and NO are as follows:

$$[N_2] = 0.036 \text{ M}; \quad [NO] = 0.017 \text{ M}.$$

Calculate the $[O_2]$.

Quiz for Chapter 14 Problems

1. When the reaction $CoCl_2 \rightleftarrows Co + Cl_2$ occurs at 25°C, the following equilibrium concentrations are observed:

$$[CoCl_2] = 3.2 \times 10^{-2} \text{ M}; \quad [Co] = 1.0 \times 10^{-5} \text{ M}; \quad [Cl_2] = 2.38 \times 10^{-4} \text{ M}$$

Calculate K_{eq} for the reaction.

2. In the equilibrium constant K_{eq}, the concentrations of the _______________ of the reaction appear in the numerator and the concentrations of the _______________ appear in the denominator.
3. The slightly soluble salt, AB, has a solubility of 5.32×10^{-6} M. Calculate the K_{sp} for AB.
4. Write the equilibrium-constant expression K_{eq} for the following reaction:

   $$2\ SO_2 + O_2 \rightleftarrows 2\ SO_3$$

5. For the reaction $X \rightleftarrows 2\ Y$, $K_{eq} = 2.3 \times 10^{-4}$. If the concentration of Y at equilibrium $= 8.5 \times 10^{-3}$, calculate the equilibrium concentration of X.
6. In a saturated solution of the slightly soluble salt CuBr, the $[Cu^+] = [Br^-] = 2.0 \times 10^{-4}$ M. Calculate the K_{sp} for CuBr.
7. For the hypothetical reaction $3\ A \rightleftarrows 2\ B$, the equilibrium concentrations are $[A] = 1 \times 10^{-3}$ M; $[B] = 2 \times 10^{-1}$ M. Calculate K_{eq} for the reaction.
8. For the reaction $2\ SO_2 + O_2 \rightleftarrows 2\ SO_3$, $K_{eq} = 810$. The equilibrium concentration of $SO_2 = 0.20$ M, and the equilibrium concentration of $SO_3 = 1.61$ M. What is the equilibrium concentration of O_2?
9. The K_{sp} value for the slightly soluble salt AgI(*s*) is 1.5×10^{-16} at 25°C. What is the solubility of AgI at 25°C?
10. If [Ag+] in a saturated solution of AgBr is 9×10^{-7} M and the [Ag+] in a saturated solution of AgCl is 1×10^{-5} M, which salt is more soluble in water? Explain.
11. The K_{eq} values for three different reactions are shown below.

    $A \rightleftarrows B$: $K_{eq} = 1 \times 10^{-2}$
    $E \rightleftarrows F$: $K_{eq} = 1 \times 10^{-18}$
    $X \rightleftarrows Y$: $K_{eq} = 1 \times 10^{5}$

    Which reaction will contain more reactant in its equilibrium state? Explain.
12. Write the K_{eq} expression for the following reaction:

    $$NH_4HS(s) \rightleftarrows NH_3(g) + H_2S(g)$$

CHAPTER

15

Electrolytes, Acids, and Bases

Overview: What You Should Be Able to Do

Chapter 15 provides a discussion of the properties and reactions of acids, bases, and salts. These compounds are known as electrolytes because solutions made with them conduct electricity. After mastering Chapter 15, you should be able to solve the following types of problems:

1. Classify a given compound as a strong acid, weak acid, strong base, weak base, or salt, and determine whether the compound is a strong electrolyte, a weak electrolyte, or a nonelectrolyte.
2. Write the formula of the salt that is produced from the reaction of any given acid and base.
3. Identify Brønsted–Lowry conjugate acid–base pairs.
4. Use the K_w to solve for the $[OH^-]$ when given the $[H_3O^+]$, or the $[H_3O^+]$ when given the $[OH^-]$.
5. Write the reaction that occurs when strong acid or strong base is added to a buffered solution.

Chart 15.1 Classifying a given compound as a strong acid, weak acid, strong base, weak base, or salt, and determining whether the compound is a strong electrolyte, a weak electrolyte, or a nonelectrolyte

Step 1

- Determine whether the compound is an acid, base, salt, or neither.
- An acid has the general formula HX (subscripts not included), where X is a monatomic or polyatomic ion.

- Many bases have the general formula $M(OH)_n$, where M is a metal ion.
- NH_3 is a common weak base. Other weak bases are derivatives of NH_3 in which one or more of the H's on NH_3 have been replaced by other atoms or groups of atoms (e.g., CH_3NH_2).
- If the compound is an acid (HX), proceed to step 2.
- If the compound is a base ($M(OH)_n$) or an NH_3-type compound, proceed to step 3.
- If the compound is not an acid or base, and contains positive and negative ions, it is a salt. Salts are strong electrolytes.
- If the compound is not an acid, base, or salt, and is formed from nonmetals it is most probably a nonelectrolyte.

⬇

Step 2

- Determine whether the acid is one of the strong acids: HCl, HBr, HI, HNO_3, H_2SO_4, $HClO_4$. Strong acids are strong electrolytes.
- If the acid is not one of the strong acids, it is a weak acid. Weak acids are weak electrolytes.

⬇

Step 3

- Determine whether the base is one of the strong bases: hydroxides of the Group IA and IIA metals. Strong bases are strong electrolytes.
- If the base is not one of the strong bases, it is a weak base. Weak bases are weak electrolytes.
- Remember that NH_3 and NH_3 derivatives are weak bases.

Example 1

Classify $Ba(OH)_2$ as a strong acid, weak acid, strong base, weak base, or salt, and determine whether the compound is a strong electrolyte, a weak electrolyte, or a nonelectrolyte.

Solution

Perform step 1:

$Ba(OH)_2$ is a base because it has the general formula of a base, $M(OH)_n$.

Perform step 2:

$Mg(OH)_2$ is a Group IIA hydroxide; therefore $Ba(OH)_2$ is a strong base. Therefore $Ba(OH)_2$ is a strong electrolyte because strong bases are strong electrolytes.

Example 2

Classify HF as a strong acid, weak acid, strong base, weak base, or salt, and determine whether the compound is a strong electrolyte, a weak electrolyte, or a nonelectrolyte.

Solution

Perform step 1:

HF is an acid because it has the general formula of an acid, HX.

Perform step 2:

HF is not one of the strong acids; therefore it is a weak acid. Therefore HF is a weak electrolyte because weak acids are weak electrolytes.

Example 3

Classify LiBr as a strong acid, weak acid, strong base, weak base, or salt, and determine whether the compound is a strong electrolyte, a weak electrolyte, or a nonelectrolyte.

Solution

Perform step 1:

LiBr is a salt because it has neither the general formula of an acid, HX, nor the general formula for a base, $M(OH)_n$.

Perform step 2:

LiBr is a salt, and therefore it is a strong electrolyte.

Practice Problems

15.1 Classify each of the following as a strong acid, weak acid, strong base, weak base, or salt, and determine whether the compound is a strong electrolyte, a weak electrolyte, or a nonelectrolyte.

(a) HBr (b) H_2CO_3 (c) $HC_2O_3O_2$

15.2 Classify each of the following as a strong acid, weak acid, strong base, weak base, or salt, and determine whether the compound is a strong electrolyte, a weak electrolyte, or a nonelectrolyte.

(a) $Ca(OH)_2$ (b) $NH_2CH_2CH_3$ (c) H_2SO_4

Chart 15.2 Writing the formula of the salt that is produced from the reaction of any given acid and base

Step 1

- Determine the anion that appears in the acid.

⬇

Step 2

Determine the cation that appears in the base.

⬇

Step 3

- Combine the cation from step 2 with the anion from step 1 to determine the formula of the salt.

Example 1

Write the formula of the salt formed from the combination of HI and $Mg(OH)_2$.

Solution

Perform step 1:

The anion that appears in the acid is I^-.

Perform step 2:

The cation that appears in the base is Mg^{2+}.

Perform step 3:

Mg^{2+} and I^- combine to form the compound MgI_2.

Example 2

Write the formula of the salt formed from the combination of H_3PO_4 and LiOH.

Perform step 1:

The anion that appears in the acid is PO_4^{3-}.

Perform step 2:

The cation that appears in the base is Li^+.

Perform step 3:

Li^+ and PO_4^{3-} combine to form the compound Li_3PO_4.

Practice Problem

15.3 Write the formula of the salt formed from the combination of H_2SO_4 and KOH.

Chart 15.3 Identifying Brønsted–Lowry conjugate acid–base pairs

Step 1

- Identify two species whose formulas differ by only the presence of one H^+. This is a conjugate acid–base pair.

⬇

Step 2

- Label the species with one more H^+ as the acid.
- Label the species with one less H^+ as the base.

Example 1

Identify the conjugate acid–base pairs in the reaction below. For each pair, indicate which species is the acid and which species is the base.

$$HNO_2 + H_2O \longrightarrow NO_2^- + H_3O^+$$

Solution

Perform step 1:

HNO_2 and NO_2^- differ by only one H^+.
H_2O and H_3O^+ differ by only one H^+.
These are the two sets of conjugate acid–base pairs.

Perform step 2:

In the HNO_2 and NO_2^- pair:

HNO_2 has one more H^+ and is therefore the acid of the pair.
NO_2^- has one less H^+ and is therefore the base of the pair.

In the H_2O and H_3O^+ pair:

H_3O^+ has one more H^+ and is therefore the acid of the pair.
H_2O has one less H^+ and is therefore the base of the pair.

Example 2

Identify the conjugate acid–base pairs in the reaction below. For each pair, indicate which species is the acid and which species is the base.

$$HCl + S^{2-} \longrightarrow HS^- + Cl^-$$

Solution

Perform step 1:

HCl and Cl^- differ by only one H^+.
S^{2-} and HS^- differ by only one H^+.
These are the two sets of acid–base pairs.

Perform step 2:

In the HCl and Cl^- pair:

HCl has one more H^+ and is therefore the acid of the pair.
Cl^- has one less H^+ and is therefore the base of the pair.

In the S^{2-} and HS^- pair:

HS^- has one more H^+ and is therefore the acid of the pair.
S^{2-} has one less H^+ and is therefore the base of the pair.

Practice Problems

15.4 Identify the conjugate acid–base pairs in the reaction below. For each pair, indicate which species is the acid and which species is the base.

$$HBr + NH_3 \longrightarrow NH_4^+ + Br^-$$

15.5 Identify the conjugate acid–base pairs in the reaction below. For each pair, indicate which species is the acid and which species is the base.

$$CH_3NH_2 + H_2O \longrightarrow CH_3NH_3^+ + OH^-$$

Chart 15.4 Using the K_w to solve for the $[OH^-]$ when given the $[H_3O^+]$, or the $[H_3O^+]$ when given the $[OH^-]$

Step 1

- If given $[H_3O^+]$, plug the $[H_3O^+]$ into the K_w expression:

$$K_w = 10^{-14} = [H_3O^+] \times [OH^-]$$

or

- If given $[OH^-]$, plug the $[OH^-]$ into the K_w expression:

$$K_w = 10^{-14} = [H_3O^+] \times [OH^-]$$

⬇

Step 2

- Divide both sides of the expression by $[H_3O^+]$ to isolate $[OH^-]$.

or

- Divide both sides of the expression by $[OH^-]$ to isolate $[H_3O^+]$.

↓

Step 3

- Solve for $[OH^-]$.

or

- Solve for $[H_3O^+]$.

Example 1

Calculate the $[OH^-]$ in a solution in which $[H_3O^+] = 2.3 \times 10^{-4}$ M.

Solution

Perform step 1:

$K_w = 10^{-14} = [H_3O^+] \times [OH^-]$

$10^{-14} = 2.3 \times 10^{-4} \times [OH^-]$

Perform step 2:

$$\frac{10^{-14}}{2.3 \times 10^{-4}} = \frac{2.3 \times 10^{-4} \times [OH^-]}{2.3 \times 10^{-4}}$$

Perform step 3:

4.3×10^{-11} M $= [OH^-]$

Example 2

Calculate the $[H_3O^+]$ in a solution in which $[OH^-] = 2.7 \times 10^{-9}$.

Solution

Perform step 1:

$K_w = 10^{-14} = [H_3O^+] \times [OH^-]$

$10^{-14} = [H_3O^+] \times 2.7 \times 10^{-9}$

Perform step 1:

$$\frac{10^{-14}}{2.7 \times 10^{-9}} = \frac{2.7 \times 10^{-4} \times [OH^-]}{2.7 \times 10^{-9}}$$

Perform step 3:

3.7×10^{-6} M $= [H_3O^+]$

Practice Problems

15.6 Calculate the $[H_3O^+]$ in a solution in which $[OH^-] = 1.3 \times 10^{-2}$ M.

15.7 Calculate the $[OH^-]$ in a solution in which $[H_3O^+] = 5.8 \times 10^{-8}$ M.

Chart 15.5 Writing the reaction that occurs when strong acid (H_3O^+) or strong base (OH^-) is added to a buffered solution

Step 1

- Identify the acid and base components of the buffered solution. (The acid component is the species with one more H^+.)
- If acid (H_3O^+) is added to the buffered solution, go to step 2.
- If base (OH^-) is added to the buffered solution, go to step 3.

⬇

Step 2

- React the added acid (H_3O^+) with the base component of the buffered solution. (The H_3O^+ will add an H^+ to the base component of the buffer, producing water and the acid component of the buffer.)

⬇

Step 3

- React the added base (OH^-) with the acid component of the buffered solution. (The OH^- will take an H^+ from the acid component of the buffer, producing water and the base component of the buffer.)

Example 1

A buffered solution is prepared using HF and F^-. Write the reaction that occurs when strong acid is added to this buffered solution.

Solution

Perform step 1:

The acid component of the buffered solution is HF; the base component is F^-. Because acid is being added to the solution, we proceed to step 2.

Perform step 2:

The added H_3O^+ will react with the F^- according to the following equation:

$$F^- + H_3O^+ \longrightarrow HF + H_2O$$

Example 2

A buffered solution is prepared using H_2CO_3 and HCO_3^-. Write the reaction that occurs when strong base is added to this buffered solution.

Solution

Perform step 1:

The acid component of the buffered solution is H_2CO_3; the base component is HCO_3^-. Because base is being added to the solution, we proceed to step 3.

Perform step 3:

The added OH^- will react with the H_2CO_3 according to the following equation:

$$H_2CO_3 + OH^- \longrightarrow HCO_3^- + H_2O$$

Practice Problems

15.8 A buffered solution is prepared using H_3BO_3 and $H_2BO_3^-$. Write the reaction that occurs when strong acid is added to this buffered solution.

15.9 A buffered solution is prepared using HCN and CN^-. Write the reaction that occurs when strong base is added to this buffered solution.

Quiz for Chapter 15 Problems

1. Classify the following compounds as electrolytes or nonelectrolytes:

   (a) HBr (b) CH_3CH_3 (c) NH_3

2. Write the formula of the salt produced in the reaction of the following acids and bases:

   (a) $H_3PO_4 + NaOH$ (b) $HNO_3 + Mg(OH)_2$

3. Write the formula for the conjugate base for the following Brønsted–Lowry acids:

   (a) H_3BO_3 (b) $HC_2H_3O_2$ (c) NH_4^+

4. Classify the following as strong acids or weak acids:

   (a) HCl (b) HF (c) HI (d) HBr

5. Complete the following sentence:

   Of the species NaF, HF, NH_3, and $HC_2H_3O_2$, the strong electrolyte is ______________.

6. Calculate the $[OH^-]$ and $[H_3O^+]$ in a solution in which the ions have equal concentration.

7. Classify the following as strong bases or weak bases:

   (a) NH_3 (b) CH_3NH_2 (c) LiOH (d) $Ca(OH)_2$

8. Write the formula of the conjugate acid for the following Brønsted–Lowry bases:

   (a) NH_3 (b) NO_3^- (c) HPO_4^{2-}

9. Calculate the $[OH^-]$ in a solution in which the $[H_3O^+] = 8.9 \times 10^{-5}$ M.

10. Indicate which of the following species represent Brønsted–Lowry acid–base pairs.

    (a) HNO_3 (b) $H_2PO_4^-$ (c) H_2SO_4 (d) NO_3^-
    (e) NH_4^+ (f) H_3PO_4 (g) NH_3 (h) HSO_4^-

11. A buffered solution is made up of H_2S and HS^-.

    (a) Write the reaction that occurs when strong acid is added to the solution.

    (b) Write the reaction that occurs when strong base is added to the solution.

12. Calculate the $[H_3O^+]$ in a solution in which the $[OH^-] = 1 \times 10^{-5}$ M.

Cumulative Quiz for Chapters 13, 14, and 15

1. If a reaction container feels hot to the touch as the reaction proceeds, is the reaction endothermic or exothermic? Explain.
2. The reaction producing NOBr from NO and Br_2 is shown in the unbalanced equation below.

   $$NO_2 + Br_2 \longrightarrow NOBr + O_2$$

   The reaction is first-order with respect to Br_2 and second-order with respect to NO_2.

   (a) Write the rate law for the reaction.

   (b) What is the overall order of the reaction?

   (c) What happens to the reaction rate when you double the Br_2 concentration?

   (d) What happens to the reaction rate when you triple the NO_2 concentration?
3. Draw a reaction-energy profile diagram for a hypothetical chemical reaction with the following characteristics. The energy of the reactants is 40 kJ/mol, the energy of the products is 20 kJ/mol, and the energy of the transition state is 60 kJ/mol. Is the reaction endothermic or exothermic? Explain.
4. Write the general rate law for the following reactions, using x and y as orders:

   $$CH_3Br + OH^- \longrightarrow CH_3OH + Br^-$$

5. A reaction consumes 250 kJ of energy.

   (a) Is the reaction endothermic or exothermic?

   (b) In the reaction-energy profile, are the reactants or products higher? Explain.
6. The slow step in the mechanism of the reaction of A_2 with BC to produce AC and B_2 is shown below. Write the rate law for the reaction.

   $$A_2 + BC \longrightarrow AC + AB$$

7. For a reaction with a K_{eq} of 1×10^9, the equilibrium lies far to the __________.
8. When a system with a very small K_{eq} value reaches equilibrium, there will be many more ____________ than ____________ in the equilibrium mixture.

9. Is the following statement true or false? Explain your answer.

   Changing the initial concentrations of the reactants will change the value of K_{eq}.

10. AgCl has a solubility of 1.26×10^{-5} M. Calculate the K_{sp} for AgCl.

11. Write equilibrium-constant expressions for the following reactions:

    (a) $2\,C_2H_4(g) + 2\,H_2O(g) \rightleftharpoons 2\,C_2H_6 + O_2(g)$

    (b) $3\,NO(g) \rightleftharpoons N_2O + NO_2(g)$

12. Calculate the value of K_{eq} for the reaction $2\,NH_3(g) \rightleftharpoons N_2(g) + 3\,H_2(g)$ if the equilibrium concentrations are as follows:

    $[NH_3] = 0.0059$ M $\quad [N_2] = 0.015$ M $\quad [H_2] = 0.032$ M

13. Write the K_{eq} expression for the reaction shown below.

    $$Fe_2O_3(s) + 3\,CO(g) \rightleftharpoons 2\,Fe(l) + 3\,CO_2(g)$$

14. For the reaction $X_2 + Y_2 \rightleftharpoons 2\,XY$, the $K_{eq} = 5.32 \times 10^{-2}$ and the equilibrium concentrations of X_2 and XY are as follows:

    $[X_2] = 0.428$ M; $[XY] = 0.24$ M.

    Calculate the $[Y_2]$.

15. Write the K_{sp} expression for each of the following salts:

    (a) $CaCO_3$ (b) $Mg(OH)_2$

16. A saturated solution of $Mg(OH)_2$ has a $[OH^-]$ of 2.88×10^{-4}. Calculate $[Mg^{2+}]$.

17. The rate law for the reaction $X_2A + Y_2 \longrightarrow Y_2A + X_2$ is Rate $= k[X_2A]$. Does this rate law support a proposal that the reaction occurs via a one-step mechanism?

18. Classify each of the following substances as a strong acid, a strong base, or a salt:

    (a) $Sr(OH)_2$ (b) H_2SO_4 (c) NH_4Cl (d) HCl

19. Identify the conjugate acid–base pairs in the reaction below. For each pair, indicate which species is the acid and which species is the base.

    $$HNO_3 + NH_3 \longrightarrow NO_3^- + NH_4^+$$

20. Calculate the $[H_3O^+]$ in a solution in which $[OH^-] = 7.6 \times 10^{-5}$.

CHAPTER

16

Nuclear Chemistry

Overview: What You Should Be Able to Do

Chapter 16 provides a discussion of the reactions that involve the nuclei of atoms and the energy that these reactions produce. After mastering Chapter 16, you should be able to solve the following types of problems:

1. Determine the mass defect associated with an atom.
2. Determine the binding energy per mole of nucleons and the binding energy per nucleon for an atom.
3. Use the half-life of an isotope to determine how much of the isotope will be left after a given amount of time.
4. Determine the age of objects (date objects) using radioactive objects.

Chart 16.1 Determining the mass defect associated with an atom

Step 1

- Determine the sum of the masses of the moles of protons, neutrons, and electrons in 1 mole of atoms.

⬇

Step 2

- Identify the mass of one mole of atoms of the element (given in the problem).

⬇

Step 3

- Subtract the mass of one mole of atoms of the element (step 2) from the sum of the masses of the moles of protons, neutrons, and electrons (step 1) in 1 mole of atoms.

Example 1

Calculate the mass defect of ${}^{14}_{6}C$. The mass of one mole is 14.003 24 grams.

Solution

Perform step 1:

$$6 \text{ mol protons} \times \frac{1.007\ 30 \text{ g}}{1 \text{ mol protons}} = 6.043\ 80 \text{ g}$$

$$+$$

$$8 \text{ mol neutrons} \times \frac{1.008\ 70 \text{ g}}{1 \text{ mol neutrons}} = 8.069\ 60 \text{ g}$$

$$+$$

$$8 \text{ mol electrons} \times \frac{0.000\ 55 \text{ g}}{1 \text{ mol electrons}} = \underline{0.004\ 40 \text{ g}}$$

Total mass $= 14.117\ 80$ g

Perform step 2:

The mass of ${}^{14}_{6}C$ is 14.003 24 g/mol.

Perform step 3:

The mass defect for ${}^{14}_{6}C$ is

$$14.117\ 80 \text{ g} - 14.003\ 24 \text{ g} = 0.114\ 56 \text{ g/mol.}$$

Example 2

Calculate the mass defect of one mole of ${}^{32}_{16}S$ atoms. The mass of ${}^{32}_{16}S$ is 31.972 07 g/mol.

Solution

Perform step 1:

$$16 \text{ mol protons} \times \frac{1.007\ 30 \text{ g}}{1 \text{ mol protons}} = 16.116\ 80 \text{ g}$$

$$+$$

$$16 \text{ mol neutrons} \times \frac{1.008\ 70 \text{ g}}{1 \text{ mol neutrons}} = 16.139\ 20 \text{ g}$$

$$+$$

$$16 \text{ mol electrons} \times \frac{0.000\ 55 \text{ g}}{1 \text{ mol electrons}} = \underline{0.008\ 80 \text{ g}}$$

Total mass $= 32.264\ 80$ g

Perform step 2:

The mass of $^{32}_{16}S$ is 31.972 07 g/mol.

Perform step 3:

The mass defect for $^{32}_{16}S$ is

$$32.264\ 80\ \text{g} - 31.972\ 07\ \text{g} = 0.2927\ \text{g/mol}.$$

Practice Problems

16.1 Calculate the mass defect of a mole of $^{63}_{29}Cu$. The mass of one mole of $^{63}_{29}Cu$ is 62.929 60.

16.2 Which has a greater mass defect, $^{12}_{6}C$, with a mass of 12.000 g/mol, or $^{10}_{5}B$, with a mass of 10.0129 g/mol?

Chart 16.2 Determining the binding energy per mole of nucleons and the binding energy per mole per nucleon for an atom

Step 1

- Determine the mass defect using the procedure in Chart 16.1.

⬇

Step 2

- Use $E = mc^2$ to calculate the binding energy per mole of nucleons.
- The mass must be in units of kg, and c must be in units of m/s.
- The units on the answer will be joules, J, because $(1\ \text{kg} \cdot \text{m}^2)/(\text{s}^2 \cdot \text{mol}) = 1$ joule.

⬇

Step 3

- Calculate the binding energy per mole per nucleon by dividing the binding energy per mole of nucleons by the total number of nucleons in the atom.

Example 1

Calculate the binding energy of $^{14}_{6}C$. The mass of one mole is 14.003 24 grams. What is the binding energy per nucleon for this atom?

Solution

Perform step 1:

The mass defect of $^{14}_{6}C$ is 0.1146 g/mol. (See Chart 16.1.)

Perform step 2:

Binding energy: $E = mc^2$

$$= (0.000\,114\,6 \text{ kg/mol}) \times (3.00 \times 10^8 \text{ m/s})^2$$
$$= 1.03 \times 10^{13} (\text{kg} \cdot \text{m}^2)/(\text{s}^2 \cdot \text{mol})$$
$$= 1.03 \times 10^{13} \text{ J/mol}$$
$$= 1.03 \times 10^{10} \text{ kJ/mol}$$

Perform step 3:

$$\text{Binding energy per nucleon} = \frac{1.03 \times 10^{10} \text{ kJ / mol}}{14 \text{ nucleons}} = 7.36 \times 10^8 \text{ kJ / mol / nucleon}$$

Example 2

Calculate the binding energy per nucleon of $^{63}_{29}Cu$. The mass of one mole is 62.939 60 grams.

Solution

Perform step 1:

The mass defect of $^{63}_{29}Cu$ is calculated to be 0.5838 g/mol. (The procedures outlined in Chart 16.1 were used to get this number. You should go through the steps yourself to calculate this result.)

Perform step 2:

Binding energy: $E = mc^2$

$$= (0.000\,583\,8 \text{ kg/mol}) \times (3.00 \times 10^8 \text{ m/s})^2$$
$$= 5.25 \times 10^{13} (\text{kg} \cdot \text{m}^2)/(\text{s}^2 \cdot \text{mol})$$
$$= 5.25 \times 10^{13} \text{ J/mol}$$
$$= 5.25 \times 10^{10} \text{ kJ/mol}$$

Perform step 3:

$$\text{Binding energy per nucleon} = \frac{5.25 \times 10^{10} \text{ kJ / mol}}{63 \text{ nucleons}} = 8.33 \times 10^8 \text{ kJ / mol / nucleon}$$

Practice Problems

16.3 Calculate the binding energy per nucleon of $^{19}_{9}F$. The mass of one mole is 18.9984 grams.

16.4 Calculate the binding energy per nucleon of $^{16}_{8}O$. The mass of one mole is 15.9949 grams.

Chart 16.3 Using the half-life of an isotope to determine how much of the isotope will be left after a given amount of time

Step 1

- Use the half-life to construct a table of the amount remaining after successive half-lives.

⬇

Step 2

- Use the table constructed in step 1 to determine how much of the substance will be left after a given amount of time.
- Values in the table will also allow you to determine how long it will take for a given amount of the isotope to decay to a specified amount.

Example 1

The half-life of iodine-123 is 13.1 hours. How much of a 25.00-g sample will remain after 65.5 hours?

Solution

Perform step 1:

Construct the table of amount remaining after successive half-lives.

Half-lives elapsed	Time elapsed (in hours)	Amount remaining (in grams)
0	0.0	25.00
1	13.1	12.50
2	26.2	6.25
3	39.3	3.13
4	52.4	1.57
5	65.5	0.781

Perform step 2:

Identify the amount remaining after the specified time has elapsed. After 65.5 hours, 0.781 grams will remain.

Example 2

The half-life of phosphorus-32 is approximately 14 days. How much of a 100.00-g sample will remain after 42 days?

Solution

Perform step 1:

Construct the table of amount remaining after successive half-lives.

Half-lives elapsed	Time elapsed (in days)	Amount remaining (in grams)
0	0	100.00
1	14	50.00
2	28	25.00
3	42	12.5

Perform step 2:

Identify the amount remaining after the specified time has elapsed. After 42 days, 12.5 grams will remain.

Example 3

The half-life of $^{55}_{27}Co$ is approximately 17.5 hours. How long will it take for a 1500.0-g sample of the isotope to decay to approximately 50 g?

Solution

Perform step 1:

Construct the table to indicate the amount remaining after successive half-lives.

Half-lives elapsed	Time elapsed (in hours)	Amount remaining (in grams)
0	0	1500.0
1	17.5	750.0
2	35.0	375.0
3	52.5	187.5
4	70.0	93.8
5	87.5	46.9

Perform step 2:

It will take about 87.5 hours (five half-lives) for the isotope to decay to approximately 50 g.

Practice Problems

16.5 The half-life of $^{60}_{27}Co$ is approximately 15.27 years. How much of a 100.00-g sample will remain after 45.8 years?

16.6 $^{19}_{8}O$ has a half-life of approximately 27 seconds. Approximately how long will it take for 250 grams of the isotope to decay to 4 grams?

Chart 16.4 Determining the age of objects (dating objects) using radioactive objects

Step 1

- Use the formula shown below to calculate the age of the object.

$$\text{Age} = \frac{-2.303 \times \log(\%\ \text{radioactive isotope remaining} / 100)}{0.693} \times \text{half-life}$$

⬇

Step 2

- Substitute the given values into the equation and solve for the unknown variable.

Example 1

$^{238}_{92}U$ has a half-life of 4.46×10^9 years as it decays to $^{206}_{82}Pb$. If a rock contains 45 atoms of $^{238}_{92}U$ for every 55 atoms of $^{206}_{82}Pb$, how old is the rock? (Note that the total number of atoms is 100. This means that the % radioactive isotope remaining is 45%.)

Solution

Perform step 1:

$$\text{Age} = \frac{-2.303 \times \log(\%\ \text{radioactive isotope remaining} / 100)}{0.693} \times \text{half-life}$$

Perform step 2:

The equation becomes

$$\text{Age} = \frac{-2.303 \times \log(45 / 100)}{0.693} \times 4.46 \times 10^9 = 5.14 \times 10^9 \text{ years}$$

Example 2

$^{14}_{6}C$ has a half-life of 5.715×10^3 years as it decays to $^{14}_{7}N$. If a cloth contains 20 atoms of $^{14}_{6}C$ for every 80 atoms of $^{14}_{7}N$, how old is the cloth?

Solution

Perform step 1:

$$\text{Age} = \frac{-2.303 \times \log(\%\ \text{radioactive isotope remaining} / 100)}{0.693} \times \text{half-life}$$

Perform step 2:

The equation becomes

$$\text{Age} = \frac{-2.303 \times \log(20/100)}{0.693} \times 5.715 \times 10^3 = 1.33 \times 10^4 \text{ years}$$

Practice Problems

16.7 $^{238}_{92}U$ has a half-life of 4.46×10^9 years as it decays to $^{206}_{82}Pb$. If a rock contains 85 atoms of $^{238}_{92}U$ for every 15 atoms of $^{206}_{82}Pb$, how old is the rock?

16.8 $^{14}_{6}C$ has a half-life of 5.715×10^3 years as it decays to $^{14}_{7}N$. If a fossil contains 5 atoms of $^{14}_{6}C$ for every 95 atoms of $^{14}_{7}N$, how old is the fossil?

Quiz for Chapter 16 Problems

1. $^{14}_{6}C$ has a half-life of 5.715×10^3 years as it decays to $^{14}_{7}N$. If a scroll contains 10 atoms of $^{14}_{6}C$ for every 90 atoms of $^{14}_{7}N$, how old is the scroll?
2. $^{238}_{92}U$ has a half-life of 4.46×10^9 years as it decays to $^{206}_{82}Pb$. If a rock contains 35 atoms of $^{238}_{92}U$ for every 65 atoms of $^{206}_{82}Pb$, how old is the rock?
3. Calculate the mass defect of $^{9}_{4}Be$. The mass of one mole is 9.0122 grams.
4. Calculate the binding energy per nucleon for $^{9}_{4}Be$. Use the result from problem 3 to perform the calculation.
5. Calculate the binding energy per nucleon of $^{40}_{20}Ca$. The mass of one mole is 39.9626 grams.
6. The half-life of $^{207}_{83}Bi$ is 35 years. How much of a 500.0-g sample of this isotope of bismuth remains after 175 years?
7. The half-life of $^{93}_{39}Y$ is approximately 10 hours. How long will it take for a 25-g sample of the isotope to decay to approximately 3 g?
8. Determine which has a greater mass defect per nucleon: $^{40}_{18}Ar$, with a mass of 39.9624 g/mol, or $^{27}_{13}Al$, with a mass of 26.9815 g/mol.
9. The difference between the sum of the masses of the atomic particles in an atom and the mass of the atom is called the __________ __________.
10. What fraction of a radioactive isotope will remain after 200 years if the isotope has a half-life of 25 years?

CHAPTER

17

The Chemistry of Carbon

Overview: What You Should Be Able to Do

Chapter 17 presents an overview of the field of organic chemistry, the chemistry of carbon compounds. After mastering Chapter 17, you should be able to solve the following types of problems:

1. Determine whether a hydrocarbon is saturated or unsaturated.
2. Name branched hydrocarbons.
3. Classify organic compounds according to the functional group present in the structure of the compound.

Chart 17.1 Determining whether a hydrocarbon is saturated or unsaturated

Step 1

- Draw the structure of the saturated hydrocarbon with the number of carbons specified.

⬇

Step 2

- Determine whether the hydrocarbon in question has the same number or a smaller number of hydrogens than the saturated hydrocarbon.
- If the number of hydrogens is the same as in the structure of the saturated hydrocarbon, the hydrocarbon in question is saturated.
- If the number of hydrogens is less than the number of hydrogens in the saturated hydrocarbon, the hydrocarbon in question is unsaturated.

⬇

Step 3

- Use the C_nH_{2n+2} general formula for saturated hydrocarbons to check the result obtained in step 2.

Example 1

The hydrocarbon that is commonly used to make film for packaging has the formula C_3H_6. Is this hydrocarbon saturated or unsaturated?

Solution

Perform step 1:

The structure of the saturated hydrocarbon containing three carbons is

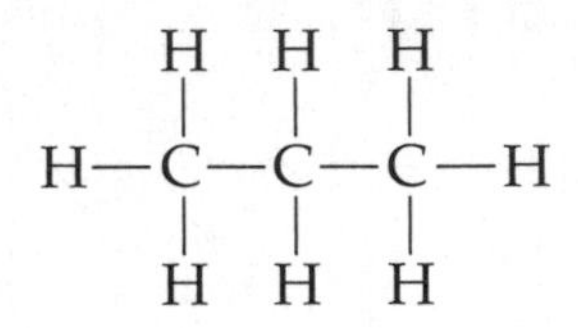

Perform step 2:

The saturated hydrocarbon has eight hydrogens, and the hydrocarbon in question has only six hydrogens. Therefore C_3H_6 is unsaturated.

Perform step 3

For a three-carbon hydocarbon, $C_nH_{2n+2} = C_3H_8$. This verifies that a saturated hydrocarbon with three carbons would have eight hydrogens; one with only six hydrogens is unsaturated.

Example 2

The hydrocarbon that is commonly used to make synthetic rubber has the formula C_4H_6. Is this hydrocarbon saturated or unsaturated?

Solution

Perform step 1:

The structure of the saturated hydrocarbon containing four carbons is

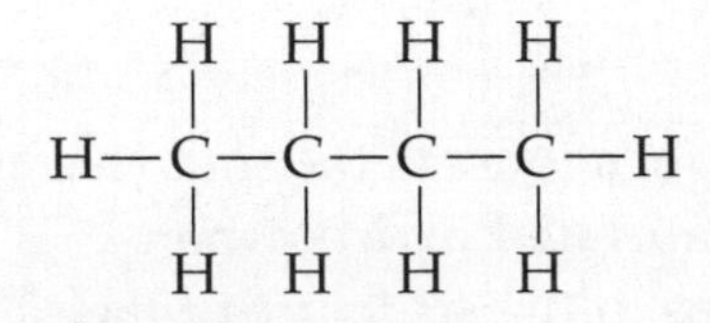

Perform step 2:

The saturated hydrocarbon has ten hydrogens, and the hydrocarbon in question has only six hydrogens. Therefore C_4H_6 is unsaturated.

Perform step 3:

For a four-carbon hydrocarbon, $C_nH_{2n+2} = C_4H_{10}$. This verifies that a saturated hydrocarbon with four carbons would have ten hydrogens; one with only six hydrogens is unsaturated.

Practice Problems

17.1 Is C_8H_{18} a saturated or unsaturated hydrocarbon?

17.2 Is C_6H_{12} a saturated or unsaturated hydrocarbon?

Chart 17.2 Naming branched hydrocarbons

Step 1

- Identify, circle, and name the main chain (longest continuous chain) in the molecule. The ending will be *-ane* if it is an alkane (single bonds only), *-ene* if an alkene (one or more double bonds), and *-yne* if an alkyne (one or more triple bonds).

⬇

Step 2

- Identify and name the branches off of the main chain. Name the branches according to the number of carbons in each branch.

⬇

Step 3

- Give the location of each branch on the main chain.
- Number the carbons in the longest chain in such a way as to give the lowest combination of numbers for the branches off of the main chain.
- If more than one branch with the same structure is present, use the prefixes *di-*, *tri-*, *tetra-*, *penta-*, and so on. and give the number of the position of each of the branches on the chain.
- In an alkene or alkyne, the longest continuous chain is numbered from the end that gives the lowest number(s) for the position of the multiple bond(s). The number of the carbon that begins the multiple bond must be specified.

Example 1

Name the following compound:

$$CH_3CH_2CH_2CH_2CH_2\underset{\underset{\displaystyle CH_3}{|}}{\underset{\displaystyle CH_2}{\underset{|}{C}}}HCH_2CH_3$$

Solution

Perform step 1:

The longest chain consists of eight carbons and is therefore named octane.

Perform step 2:

The branch is named ethyl because it has two carbons in it.

Perform step 3:

The chain must be numbered from the right, in order to give the branch the position of 3 and not 6. Thus, the branch is named 3-ethyl. The compound is therefore named 3-ethyloctane.

Example 2

Name the following compound:

$$\underbrace{HC{\equiv}C{-}\underset{\displaystyle CH_3}{\underset{|}{C}}HCH_3}$$

Solution

The longest chain consists of four carbons and is therefore named butyne (ending in *-yne* because of the triple bond).

Perform step 2:

The branch is named methyl because it has one carbon in it.

Perform step 3:

The chain must be numbered from the left in order to give the triple bond the lowest numbered position (1-butyne). Thus, the branch is named 3-methyl. The compound is therefore named 3-methyl-1-butyne.

Example 3

Name the following compound:

$$CH_2{=}CH{-}\underset{\displaystyle \begin{array}{c} CH_2 \\ | \\ CH_2 \\ | \\ CH_3 \end{array}}{\underset{|}{C}H}{-}CH_3$$

Solution

Perform step 1:

The longest chain consists of six carbons and is therefore named hexene (ending in *-ene* because of the double bond).

Perform step 2:

The branch is named methyl because it has one carbon in it.

Perform step 3:

The chain must be numbered from the left in order to give the double bond the lowest numbered position (1-hexene). Thus, the branch is named 3-methyl. The compound is therefore named 3-methyl-1-hexene.

Practice Problems

17.3 Name the following compound:

$$\begin{array}{l} CH_3C \equiv \underset{\displaystyle \begin{array}{c} | \\ CH_2 \\ | \\ CH_3 \end{array}}{C}HCH_3 \end{array}$$

17.4 Name the following compound:

$$CH_3CH \equiv CH\underset{\displaystyle \begin{array}{c} | \\ CH_3 \end{array}}{C}HCH_2CH_3$$

Chart 17.3 Classifying organic compounds according to the functional group present in the structure of the compound

Step 1

- Use the chart of classes of organic compounds to locate the functional group found in the molecule.

↓

Step 2

- Classify the compound based on its functional group.
- If more than one functional group is present, the compound can generally be classified as both (e.g., $CH_2{=}CH{-}CH_2OH$ can be classified as an alkene and as an alcohol).

Example 1

Classify the following compound according to the group in which it belongs:

$$CH_3CH_2NH_2$$

Solution

Perform step 1:

According to the chart, a compound with a nitrogen bonded to two hydrogens is an amine.

Perform step 2:

The compound is an amine.

Example 2

Classify the following compound according to the group in which it belongs:

$$CH_3CH_2OCH_2CH_2CH_3$$

Solution

Perform step 1:

According to the chart, a compound with an oxygen between two alkyl groups is an ether.

Perform step 2:

The compound is an ether.

Example 3

Classify the following compound according to the group in which it belongs:

$$CH_3CH_2\overset{\overset{\displaystyle O}{\|}}{C}-H$$

Solution

Perform step 1:

According to the chart, a compound with a carbon doubly bonded to an oxygen and to a hydrogen is an aldehyde.

Perform step 2:

The compound is an aldehyde.

Practice Problems

17.5 Classify the following compound:

$$CH_3CH_2-\overset{\overset{\large O}{\|}}{C}-CH_2CH_3$$

17.6 Classify the following compound:

$$CH_3-\overset{\overset{\large O}{\|}}{C}-OH$$

17.7 Classify the following compound:

CH_3F

Quiz for Chapter 17 Problems

1. Name the following compound:

$$\begin{array}{l} CH_3CH_2\underset{\displaystyle |\atop\displaystyle CH_3}{CH}CH_2CH_3 \end{array}$$

2. Fill in the blanks in the following sentence:

   A hydrocarbon that contains only single bonds is a(n) ____________, one with a double bond is a(n) ______________, and one containing a triple bond is a(n) ____________.

3. Classify the following compounds according to the functional group to which each belongs:

   (a) $CH_3\underset{\displaystyle |\atop\displaystyle CH_3}{C}HCHOH$ (b) $CH_3CH_2OCH_2CH_3$ (c) $CH_3CH_2CH_2CH_2Br$

4. Indicate whether the following compounds are saturated or unsaturated:

   (a) C_5H_{12} (b) C_7H_{14}

5. Write the structure of the functional group for each of the following classes of compounds:

   (a) alcohols (b) ethers (c) carboxylic acids

6. Provide the name of the unbranched alkane that contains the following number of carbons in its main chain:

   (a) three carbons (b) seven carbons

7. Name the compound below.

$$\begin{array}{l}CH_3CH_2CHCH_2CH{=}CH_2\\ \qquad\quad\;\;|\\ \qquad\quad CH_3\end{array}$$

8. The three classes of compounds in which there is carbon doubly bonded to an oxygen are __________, __________, and __________.
9. A saturated hydrocarbon that contains 12 carbons will contain ______ hydrogens.
10. Draw the structure of a molecule that might be called 2-methyl-3-butene. Write the correct name of the molecule.

CHAPTER

18

Synthetic and Biological Polymers

Overview: What You Should Be Able to Do

Chapter 18 provides an introduction to the types of large organic compounds that are known as polymers. After mastering Chapter 18, you should be able to solve the following types of problems:

1. Draw the polymer formed from a given alkene monomer.
2. Draw the polymer formed from combining a carboxylic acid or acid chloride with an amine.
3. Draw the monomer(s) used to synthesize a given alkene polymer.

Chart 18.1 Drawing the polymer formed from a given alkene monomer

Step 1

- Write the alkene monomer and note the two groups connected to each carbon in the alkene.

⬇

Step 2

- Connect pairs of alkene carbons by converting double bonds to single bonds (and with the two groups attached to each carbon intact).

Example 1

Draw three repeating units of the polymer from the monomer shown below.

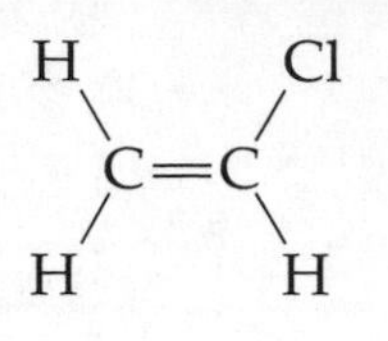

Solution

Perform step 1:

The structure of the alkene is

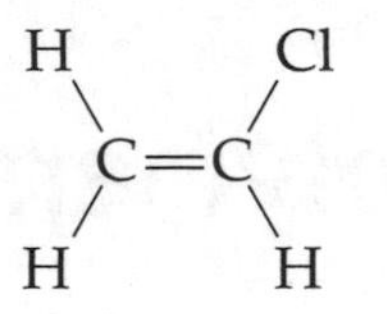

Perform step 2:

Connecting the carbons from two monomers, and converting the double bond to single bonds, we obtain the following polymer:

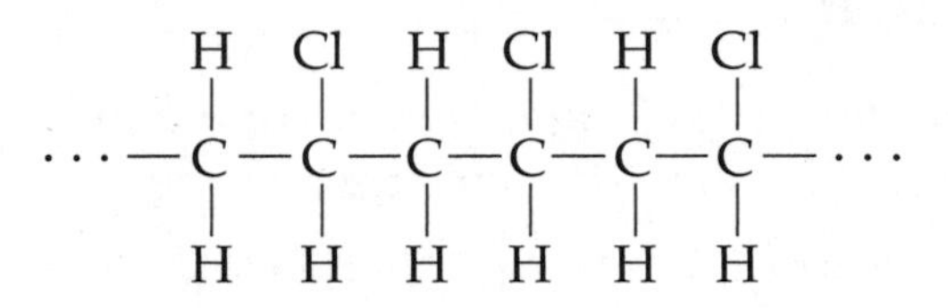

Example 2

Draw three repeating units of the polymer from the monomer shown below.

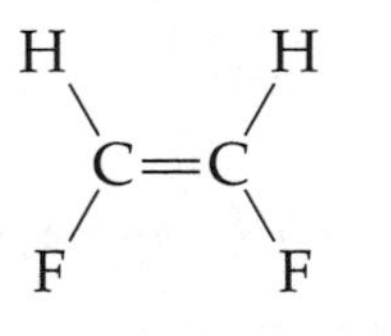

Solution

Perform step 1:

The structure of the alkene is

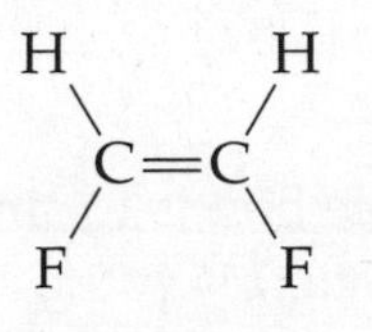

Perform step 2:

Connecting the carbons from two monomers, and converting the double bond to single bonds, we obtain the following polymer:

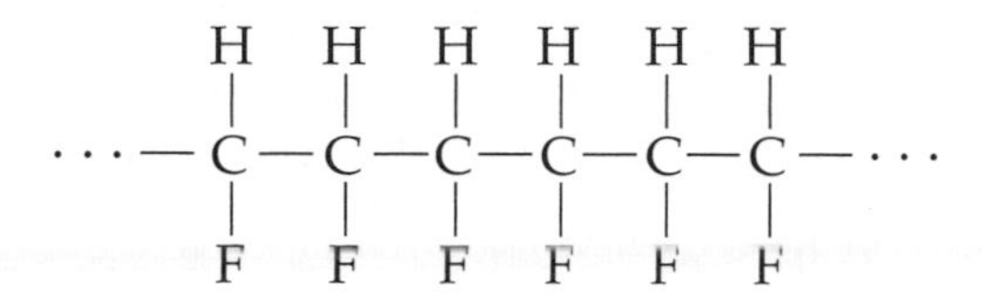

Practice Problems

18.1 Draw three repeating units of the polymer from the monomer shown below.

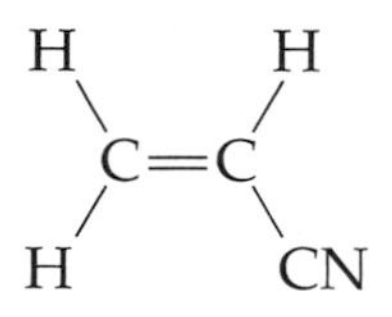

18.2 Draw three repeating units of the polymer from the monomer shown below.

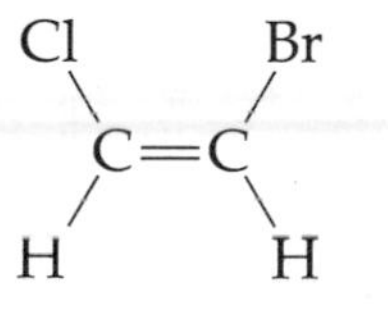

Chart 18.2 Drawing the polymer formed from combining a carboxylic acid or acid chloride with an amine

Step 1

- Remove the atom connected to the C=O (the OH in a carboxylic acid or the Cl in an acid chloride).
- Remove an H connected to the N in the amine.
- These groups combine to form water or HCl.

⬇

Step 2

- Form the amide bond by making a bond between the C from the carboxylic acid (or acid chloride) and the N from the amine.

Example 1

Draw three repeating units of the polymer formed from the combination of the following monomers:

$$H_2NCH_2NH_2 + HO\overset{O}{\overset{\|}{C}}CH_2CH_2\overset{O}{\overset{\|}{C}}OH$$

Solution

Perform step 1:

Removing the OH's from the C=O and an H from the NH_2 (to form water) we obtain

$$-\underset{H}{\underset{|}{N}}CH_2\underset{H}{\underset{|}{N}}- \quad \text{and} \quad -\overset{O}{\overset{\|}{C}}CH_2CH_2\overset{O}{\overset{\|}{C}}-$$

Perform step 2:

Connecting the N's with the C=O, we get the polymer

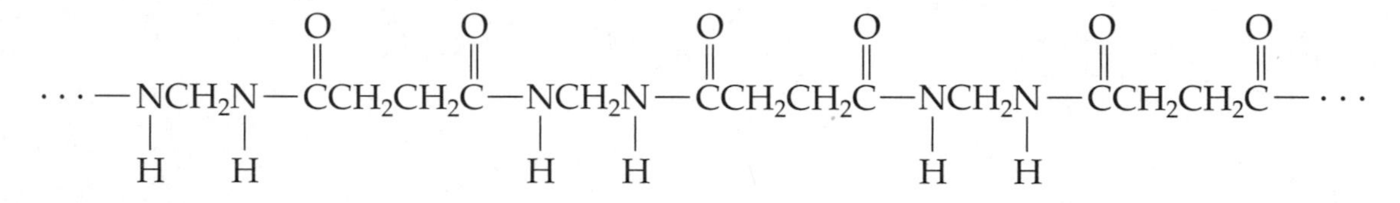

$$\cdots-\underset{H}{\underset{|}{N}}CH_2\underset{H}{\underset{|}{N}}-\overset{O}{\overset{\|}{C}}CH_2CH_2\overset{O}{\overset{\|}{C}}-\underset{H}{\underset{|}{N}}CH_2\underset{H}{\underset{|}{N}}-\overset{O}{\overset{\|}{C}}CH_2CH_2\overset{O}{\overset{\|}{C}}-\underset{H}{\underset{|}{N}}CH_2\underset{H}{\underset{|}{N}}-\overset{O}{\overset{\|}{C}}CH_2CH_2\overset{O}{\overset{\|}{C}}-\cdots$$

Example 2

Draw three repeating units of the polymer formed from the combination of the following monomers:

$$H_2NCH_2NH_2 + Cl\overset{O}{\overset{\|}{C}}CH_2CH_2\overset{O}{\overset{\|}{C}}Cl$$

Solution

Perform step 1:

Removing the Cl's from the C=O and an H from the NH_2 (to form HCl) we obtain

$$-\underset{H}{\underset{|}{N}}CH_2CH_2\underset{H}{\underset{|}{N}}- \quad \text{and} \quad -\overset{O}{\overset{\|}{C}}CH_2\overset{O}{\overset{\|}{C}}-$$

Perform step 2:

Connecting the N's with the C=O, we get the polymer

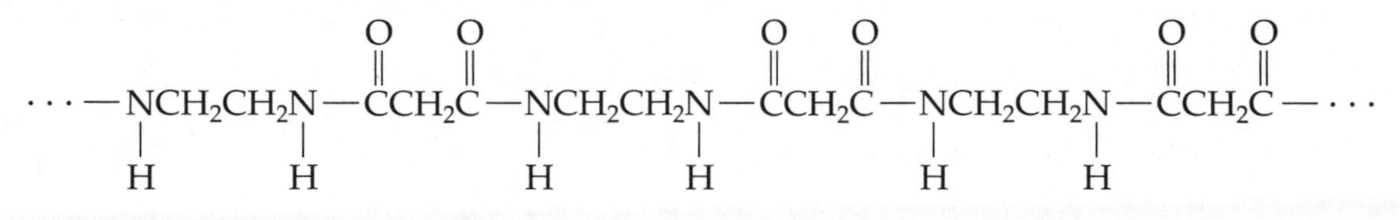

Example 3

Draw three repeating units of the polymer formed from the combination of units of the following monomer:

$$H_2NCH_2\overset{\overset{\large O}{\|}}{C}OH$$

Solution

Perform step 1:

Removing the OH's from the C=O and an H from the NH_2 we obtain

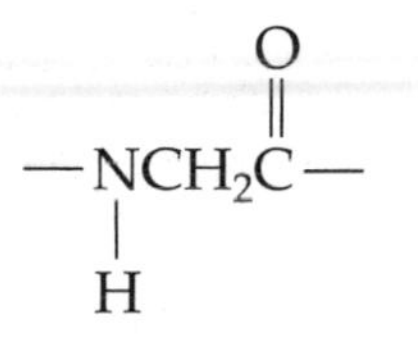

Perform step 2:

Connecting the N's with the C = O, we get the polymer

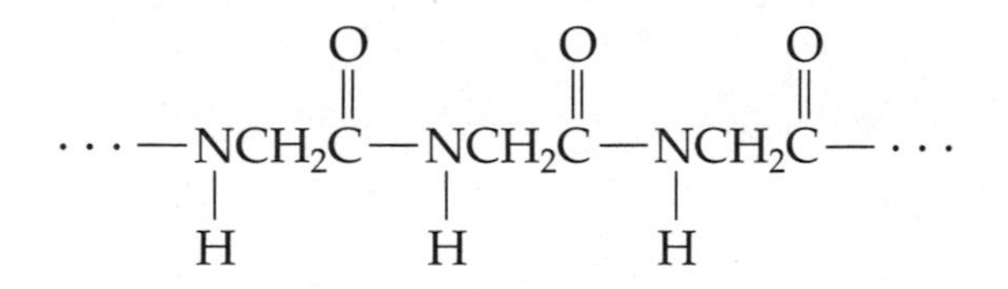

Practice Problems

18.3 Draw three repeating units of the polymer formed from the combination of units of the following monomer:

$$H_2NCH_2CH_2CH_2\overset{\overset{\large O}{\|}}{C}OH$$

18.4 Draw three repeating units of the polymer formed from the combination of the following monomers:

$$H_2NCH_2CH_2NH_2 + HO\overset{\overset{\displaystyle O}{\|}}{C}CH_2CH_2\overset{\overset{\displaystyle O}{\|}}{C}OH$$

Chart 18.3 Drawing the monomer(s) used to synthesize a given alkene polymer

Step 1

- Identify the repeating units in the polymer.

⬇

Step 2

- Draw one of these repeating units, placing a double bond between the carbon atoms.

Example 1

Write the monomer used to produce the following polymer:

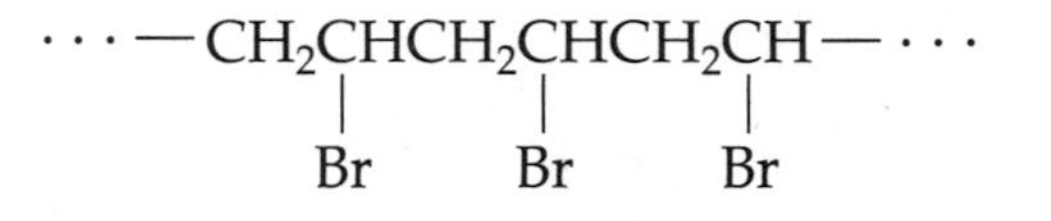

Solution

Perform step 1:

The repeating unit is

$$\cdots-CH_2\underset{\underset{\displaystyle Br}{|}}{C}H-\cdots$$

Perform step 2:

Putting a double bond between the two carbons, we obtain $CH_2{=}CHBr$ as the monomer used to produce the polymer.

Example 2

Write the monomer used to produce the following polymer:

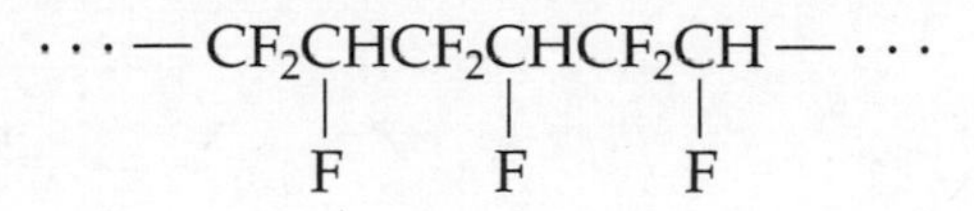

Solution

Perform step 1:

The repeating unit is

$$\cdots - \underset{\displaystyle \mathrm{F}}{\mathrm{CF_2}\underset{|}{\mathrm{CH}}} - \cdots$$

Perform step 2:

Putting a double bond between the two carbons, we obtain $CF_2{=}CHF$ as the monomer used to produce the polymer.

Practice Problems

18.5 Write the monomer used to produce the following polymer:

$$\cdots - \mathrm{CF_2}\underset{\displaystyle \mathrm{Br}}{\underset{|}{\mathrm{CH}}}\mathrm{CF_2}\underset{\displaystyle \mathrm{Br}}{\underset{|}{\mathrm{CH}}}\mathrm{CF_2}\underset{\displaystyle \mathrm{Br}}{\underset{|}{\mathrm{CH}}} - \cdots$$

18.6 Write the monomer used to produce the following polymer:

$$\cdots - \underset{\displaystyle \mathrm{Br}}{\underset{|}{\mathrm{CH}}}\underset{\displaystyle \mathrm{Br}}{\underset{|}{\mathrm{CH}}}\underset{\displaystyle \mathrm{Br}}{\underset{|}{\mathrm{CH}}}\underset{\displaystyle \mathrm{Br}}{\underset{|}{\mathrm{CH}}}\underset{\displaystyle \mathrm{Br}}{\underset{|}{\mathrm{CH}}}\underset{\displaystyle \mathrm{Br}}{\underset{|}{\mathrm{CH}}} - \cdots$$

18.7 Write the monomer used to produce the following polymer:

$$\cdots - \mathrm{CH_2}\underset{\displaystyle \mathrm{CH_3}}{\underset{|}{\mathrm{CH}}}\mathrm{CH_2}\underset{\displaystyle \mathrm{CH_3}}{\underset{|}{\mathrm{CH}}}\mathrm{CH_2}\underset{\displaystyle \mathrm{CH_3}}{\underset{|}{\mathrm{CH}}} - \cdots$$

Quiz for Chapter 18 Problems

1. Draw two repeating units of the polymer formed from the monomer below.

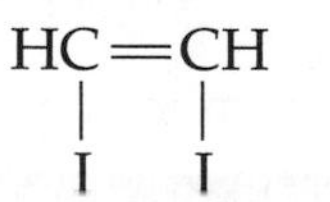

2. Draw the monomer that was used to make the following polymer.

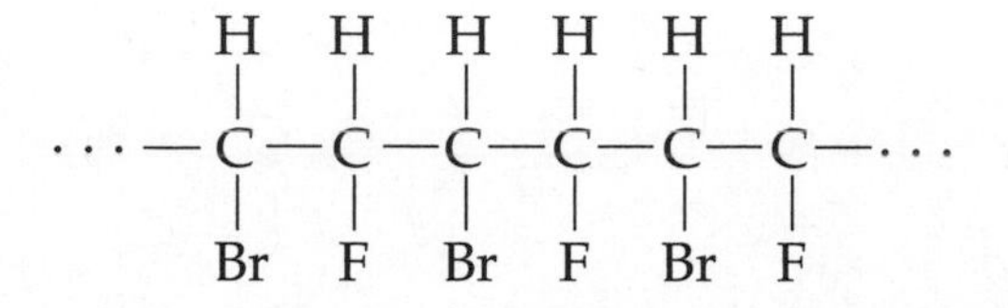

3. Fill in the blank in the following sentence.

   The bond between the carbon from a carbonyl group and the nitrogen from an amine is called a(n) ________________ bond.

4. Draw a monomer that can be used to make the following polymer:

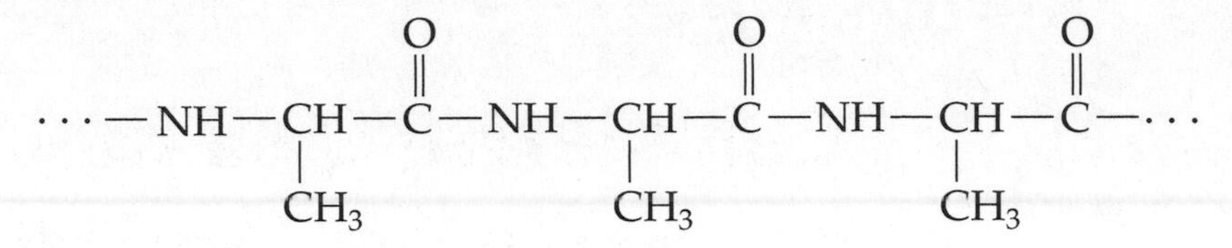

5. Fill in the blank in the following sentence:

   When a carboxylic acid combines with an amine to form a polymer, the small molecule that is eliminated in the process is ____________.

6. Alanine is one of 20 amino acids that are found in proteins. The structure of alanine is shown below.

   H_3C O
      | ‖
   $H_2NCHCOH$

   Write the structure of the polymer that would be formed from a combination of alanine units.

7. Write the structure of the monomer that would be used to produce the following polymer:

   $\cdots-CH_2Cl_2CH_2Cl_2CH_2Cl_2-\cdots$

8. Fill in the blanks in the sentence below.

   When an alkene polymer is formed from a monomer unit, the __________ in each monomer is converted to a ___________ in the polymer.

9. Draw two repeating units of the polymer that is formed from the combination of units of the monomer shown below.

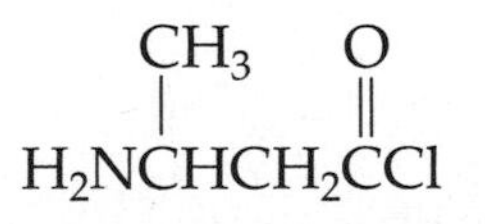

10. Complete the following sentence:

    When the polymer is formed from the monomer in question 9, the small molecule that is eliminated when the bond is formed is ________.

Cumulative Quiz for Chapters 16, 17, and 18

1. $^{14}_{6}C$ has a half-life of 5.715×10^3 years as it decays to $^{14}_{7}N$. If a fossil contains 20 atoms of $^{14}_{6}C$ for every 70 atoms of $^{14}_{7}N$, how old is the fossil?
2. Calculate the mass defect of $^{13}_{6}C$. The mass of one mole is 13.003 35 g.
3. Calculate the binding energy per nucleon of $^{13}_{6}C$. The mass of one mole is 13.003 35 g.
4. The half-life of $^{209}_{84}Po$ is 105 years. How much of a 200.0-g sample of this isotope of polonium will remain after 525 years?
5. A 5.00-kg sample of a radioactive substance was found to have decayed to 0.3125 kg in a period of 20 years. What is the half-life of the substance?
6. Name the following compound:

   $$\begin{array}{l} CH_3CH_2\underset{\displaystyle CH_3}{\underset{|}{C}}HCH_2CH_2CH_2CH_3 \end{array}$$

7. Classify the following compounds according to the functional group to which each belongs:

   (a) $CH_3\underset{\displaystyle CH_3}{\underset{|}{C}}HCH_2CH_2OH$ (b) $CH_3\overset{\displaystyle O}{\overset{\|}{C}}OH$ (c) $CH_3CH_2CH_2F$

8. Indicate whether the following compounds are saturated or unsaturated:

   (a) C_4H_8 (b) C_5H_{12}

9. Provide the name of the alkane that contains the following number of carbons in its main chain:

   (a) six carbons (b) eight carbons

10. Name the compound below.

    $$CH_3\underset{\displaystyle CH_3}{\underset{|}{C}}HCH{=}CH_2$$

11. Draw the structure of a molecule that might be called 2-methyl-3-butene. Write the correct name of the molecule.

12. Draw two repeating units of the polymer formed from the monomer below.

$$\underset{\text{Br}}{\text{CH}}{=}\underset{\text{I}}{\text{CH}}$$

13. Draw the monomer that was used to make the following polymer:

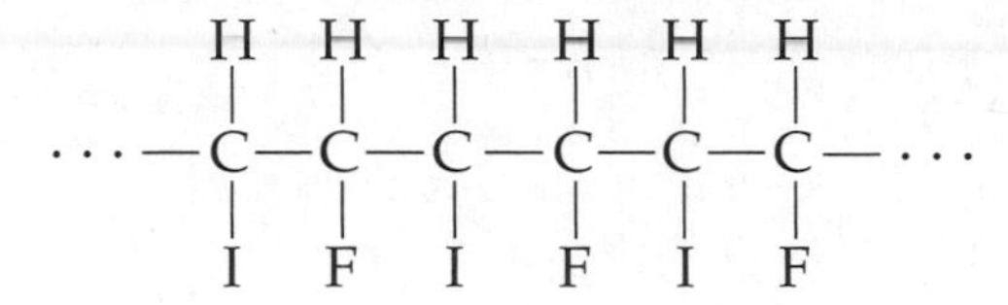

14. Fill in the blank in the following sentence:

When alkene monomers form a polymer, the _________ bonds are converted to _________ bonds.

15. Draw a monomer that can be used to make the following polymer:

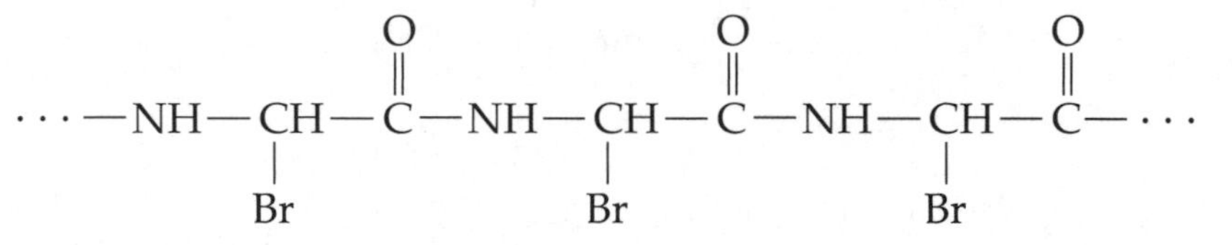

16. Write the structure of the monomer that would be used to produce the following polymer:

$$\cdots{-}CH_2CBr_2CH_2CBr_2CH_2CBr_2{-}\cdots$$

17. Draw two repeating units of the polymer that is formed from the combination of units of the monomer shown below.

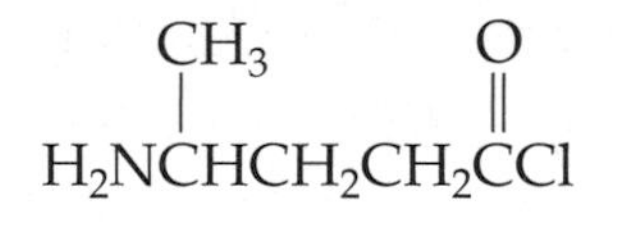

18 Draw two repeating units of the polymer formed from the combination of the following monomers:

$$H_2NCH_2CH_2NH_2 + HO\overset{O}{\overset{\|}{C}}CH_2CH_2CH_2\overset{O}{\overset{\|}{C}}OH$$

19. Write the monomer used to produce the following polymer:

$$\cdots{-}CH_2\underset{F}{CH}CH_2\underset{F}{CH}CH_2\underset{F}{CH}{-}\cdots$$

20. Write two repeating units of the polymer that would be formed from a combination of the monomer unit shown below.

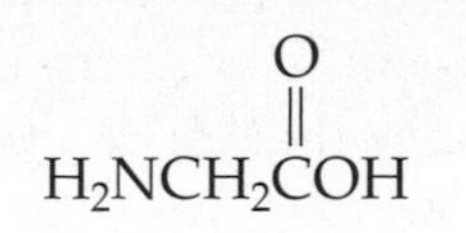

APPENDIX

Mathematics Review

Overview: What You Should Be Able to Do

1. Perform arithmetic operations on signed (posi tive and negative) numbers.
2. Solve problems involving the manipulation of fractions, decimals, and percentages.
3. Solve first-degree (linear) algebraic equations.
4. Perform calculations involving numbers expressed as exponents or roots.
5. Express numbers in scientific notation, and perform arithmetic operations on numbers written in scientific notation.
6. Determine the values of logarithms and inverse logarithms for given numbers.
7. Determine the relationship between two variables by examining a graph.

A.1 Signed Numbers

Both positive and negative numbers are used to describe the values of quantities and properties. Certain properties can have only positive values, such as the number of atoms in a molecule. Some variables, however, can have negative values, such as the temperature in degrees Fahrenheit or Celsius. Positive numbers will generally be written with no sign in front of them, but negative numbers will always have a minus sign in front of them. For example, a temperature of 20 degrees below zero Fahrenheit will be expressed as −20°F, but a temperature of 20 degrees above zero Fahrenheit will be expressed as 20°F.

In chemistry you will have to perform arithmetic operations on both positive and negative numbers. The rules for performing arithmetic operations on signed numbers are presented below. (Although you will usually perform these operations with your calculator, it is helpful to know the rules so you will not be dependent on the calculator.) Before we consider the rules, however, we must introduce one other term: *absolute value*. The absolute value of a number is the number obtained by dropping the sign of the number. Therefore, the absolute value of any number, positive or negative, is a positive number.

A. Adding Signed Numbers

1. When the signs of the numbers are the same, add the absolute values of the numbers, and assign the answer the sign of the numbers being added.
2. When the signs of the numbers are different, subtract the smaller absolute value from the larger absolute value. The sign of the answer will be the sign of the number with the larger absolute value.
3. When adding more than two signed numbers, add all the positive numbers to get one positive number, and add all the negative numbers to get one negative number. Then add the one positive number to the one negative number, using the rule for adding numbers with different signs. Or you can add two numbers at a time, using whichever rule applies.

Example A.1

Add $-8 + 9 + (-10) + 5$.

Solution

We can add the positive numbers to get $+14$, add the negative numbers to get -18, and add $+14$ and -18 to get -4 for the final answer. Or, we can add -8 to $+9$ to get $+1$, then add -10 to $+1$ to get -9, and finally $+5$ to -9 to get -4. We get the same answer using either procedure.

Practice Exercise A.1

Add $9 + (-5) + (-8) + 14$.

B. Subtracting Signed Numbers

To subtract signed numbers, we change the sign of the number being subtracted (the subtrahend) from the original number (the minuend), and use the rules of addition.

Example A.2

Perform the following subtraction:

$$-35 - (-42)$$

Solution

$-35 - (-42) = -35 + (+42) = +7$

Practice Exercise A.2

Perform the following subtraction: $58 - (-32)$.

C. Multiplying and Dividing Signed Numbers

When multiplying and dividing signed numbers, we perform the operation on the absolute value of the numbers. If both signs are the same, the answer is positive; if the two signs are different, the answer is negative. In summary:

For multiplication:	**For division:**
$(+)(+) = +$	$+/+ = +$
$(-)(-) = +$	$-/- = +$
$(+)(-) = -$	$+/- = -$
$(-)(+) = -$	$-/+ = -$

Example A.3

Perform the following operation: $(-8)(-6)$.

Solution

Since the signs are the same, the product will be positive.

$$(-8)(-6) = +48, \text{ or simply } 48$$

Practice Exercise A.3

Perform the following operation: $-64/16$.

A.2 Fractions, Decimals and Percents

Fractions, decimals, and percentages are all related to each other. Each is used to indicate some part of a whole quantity. For example, 3/4, 0.75, and 75% are all used to express a quantity representing three parts out of four. The fraction represents the number of parts of the whole (3 parts of the whole, which is divided into 4 equal sized parts).

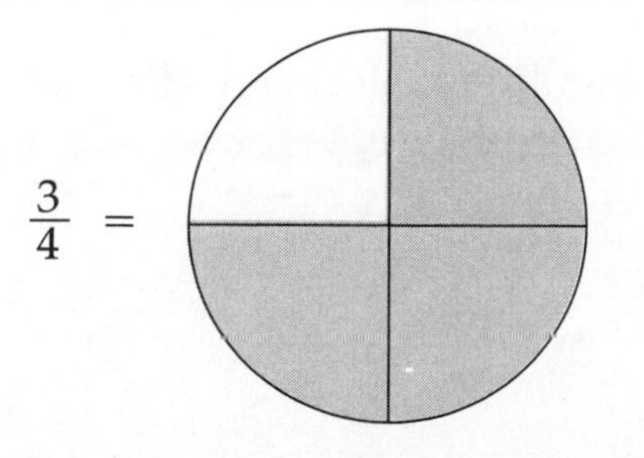

The decimal represents the part of one whole (0.75 of the whole), as shown below.

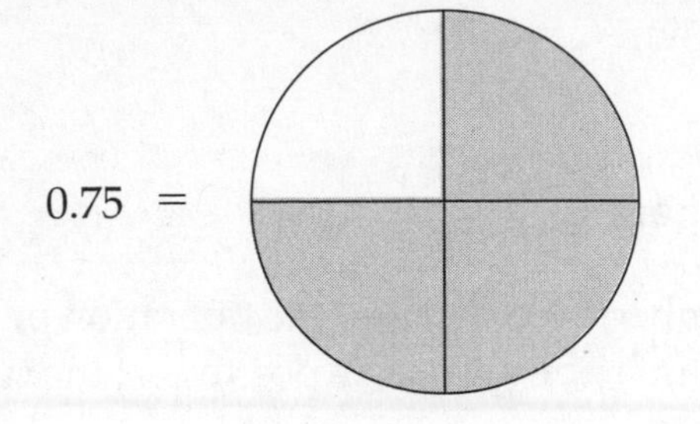

The position of a number in a decimal represents the value of the number. The values of the different positions are shown below:

tenths
hundredths
thousandths

0.750

The percentage represents the number of parts of a whole that has been divided into 100 parts (75 parts of the whole which has been divided into 100 equal parts).

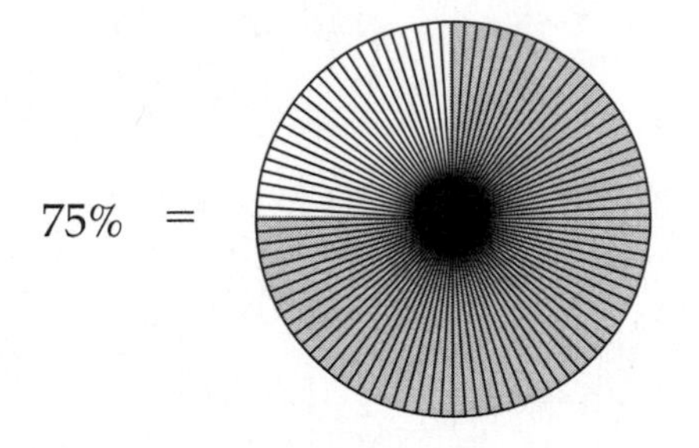

The fraction, 3/4, is converted to a decimal by dividing the numerator by the denominator, and the decimal is converted to the percentage by moving the decimal point two places to the right. (This is equivalent to multiplying the decimal by 100% in order to convert it to the percentage.) To convert a percent to a decimal, we move the decimal point two places to the left.

Example A.4

Express the fraction 3/8 as a decimal and as a percent.

Solution

To change 3/8 to a decimal, we divide the numerator (3) by the denominator (8), resulting in 0.375. To convert the decimal to a percent, we multiply the decimal by 100%, or simply move the decimal point two places to the right.

$$0.375 \times 100\% = 37.5\%$$

Practice Exercise A.4

Express the fraction 3/5 as a decimal and as a percent.

Problems involving fractions, decimals, and percents are worked the same way, the only difference being whether we perform the calculations using the fraction or the decimal form of the number. All percentages *must* be expressed as either the fraction or the decimal before any calculations are performed.

There are basically three types of fraction, decimal, or percent problems that you will encounter in your study of chemistry. Examples of the three types are presented below.

A. Problems that require you to find what percent (or what fraction) one number is of another.

In this type of problem, a percent is not given in the problem, but you are asked to calculate it.

Example A.5

What percent of 48 is 30?

Solution

To solve this type of problem, divide the part by the whole, and multiply by 100%. (We expect the answer to be greater than 50% because 24 is 50% of 48, and 30 is larger than 24.)

$$\% = \frac{\text{part}}{\text{whole}} \times 100\% = \frac{30}{48} \times 100\% = 0.625 \times 100\% = 62.5\%$$

Therefore, 30 is 62.5% of 48. (When we check for the reasonableness of our answer, we find that it meets our expectation of being greater than 50%.)

Practice Exercise A.5

What percent of 50 is 43?

Sometimes the "part" is larger than the whole, as in the following example.

Example A.6

310 is what percent of 125? (Because 310 is approximately two and one-half times as big as 125, we expect our answer to be approximately 250%.)

Solution

Because the "part" is larger than the whole, the answer will be greater than 100%, but the procedure will still be the same.

$$\% = \frac{\text{part}}{\text{whole}} \times 100\% = \frac{310}{125} \times 100\% = 2.48 \times 100\% = 248\%$$

(When we check for the reasonableness of our answer, we find that it meets our expectation of being approximately 250%.)

Practice Exercise A.6

12 is what percent of 5?

B. Problems in which you have to calculate a certain percentage of a number.

In this type of problem a percentage is given, and you are required to calculate that percentage of the given number. Example A.7 presents this type of problem.

Example A.7

What is 32% of 75?

Solution

To solve this type of problem, change the percent to a decimal (or fraction if you prefer), and multiply the decimal by the number.

$$32\% \text{ of } 75 = 0.32 \times 75 = 24$$

(Our answer is reasonable because 32% is approximately, but a little less than 1/3.)

Practice Exercise A.7

If 51% of the mass of a mixture of iron and copper is iron, how much iron is there in a sample that has a mass of 250 grams?

C. Problems in which you are required to find the whole when a percentage is given.

In this type of problem you are told the value of a certain percent of the whole, and asked to find the total, as in Example A.8 below.

Example A.8

Water consists of approximately 11% hydrogen and 89% oxygen. If a sample of water contains 300 grams of oxygen, what is the mass of the water?

Solution

Because we are given the mass of a *part* of the whole (300 grams, just the oxygen), we know that the mass of the whole will be a larger number. To work the problem we change the percent to a decimal, and then divide the given number by the decimal.

$$89\% = 0.89$$

$$\frac{300}{0.89} = 337$$

The mass of the water is therefore 337 grams, an answer that meets our expectation. Because most students find this type of problem the most difficult to "see through," you should recognize that this is the only type of problem that requires you to *divide* by a percent (changed to a decimal). This is always the procedure when the answer is larger than the number given—in other words, when we are solving for the whole having been given a part.

Practice Exercise A.8

Air is 23% oxygen (by weight). How many pounds of air would be needed in order to obtain 500 pounds of oxygen?

A.3 Linear (First-Degree) Equations

An equation is a mathematical sentence that states that two expressions are equal. One or both of the expressions will contain a variable whose value must be determined by solving the equation. Linear equations are those in which the variable being solved for is raised to a power no higher than 1. (Equations in which the variable being solved for is raised to a power of 2 are called quadratic equations. The problems in *Introductory Chemistry* do not require you to solve quadratic equations, so this topic will not be covered here.) Consider the following example of a linear equation.

Example A.9

Solve for the value of x in the equation below:

$$5x - 3 = 2x + 9$$

To solve the equation for x, the first step is to get all of the terms involving x on one side of the equation, and all other terms on the other side. We are allowed to perform the following operations on any equation without changing the equality that the equation represents.

1. Adding the same quantity to both sides of the equation.

   If $x = y$, then $x + a = y + a$

2. Subtracting the same quantity from both sides of the equation.

   If $x = y$, then $x - a = y - a$

3. Multiplying both sides of the equation by the same quantity.

   If $x = y$, then $x \cdot a = y \cdot a$

4. Dividing both sides of the equation by the same quantity.

   If $x = y$, then $\frac{x}{a} = \frac{y}{a}$

In the equation $5x - 3 = 2x + 9$, we will collect all of the x's on the left side of the equation and all of the numbers on the right side. We must add 3 to each side to get the numbers on the right side, and subtract $2x$ from each side to get the x on the left side.

$$
\begin{array}{lrcl}
\text{Original equation:} & 5x - \cancel{3} & = & 2x + 9 \\
\text{(1) Add three:} & + \cancel{3} & & + 3 \\
\hline
 & 5x & = & \cancel{2x} + 12 \\
\text{(2) Subtract } 2x\text{:} & -2x & & -\cancel{2x} \\
\hline
 & 3x & = & 12
\end{array}
$$

Now to solve for x, we must divide both sides of the equation by 3.

(3) Divide by 3: $\frac{\cancel{3}x}{\cancel{3}} = \frac{12}{3}$

Result: $x = 4$

Practice Exercise A.9

Solve for the value of x in the equation below:

$6x + 7 = 4x + 23$

Example A.10

Solve for the value of b in the equation:

$\text{rate} = k \cdot a \cdot b \cdot c$

Solution

In order to solve for b, we need to remove every other term on the same side of the equation as the b, so that the b will be isolated on one side of the equation. We will divide both sides of the equation by k, a, and c, as follows:

$$\frac{\text{rate}}{kac} = \frac{\cancel{k} \cdot \cancel{a} \cdot b \cdot \cancel{c}}{\cancel{k} \cdot \cancel{a} \cdot \cancel{c}}$$

Therefore, $\frac{\text{rate}}{kac} = b$ or $b = \frac{\text{rate}}{k \cdot a \cdot c}$

Practice Exercise A.10

Solve for the value of n in the equation:

$$PV = nRT$$

A.4 Exponents and Roots

An exponent (denoted by a superscript number) indicates the number of times the base number is multiplied by itself. For example, x^3 means the product of $x \cdot x \cdot x$, $10^4 = 10 \cdot 10 \cdot 10 \cdot 10$, and $x^1 = x$.

A negative exponent indicates the reciprocal of the quantity represented by the positive exponent of the same value—in other words, one divided by that number.

$$y^{-5} = \frac{1}{y \cdot y \cdot y \cdot y \cdot y} \text{ and } 2^{-3} = \frac{1}{2 \cdot 2 \cdot 2}$$

The rules for calculations involving exponents are as follows:

1. $x^m \cdot x^n = x^{m+n}$

   example: $10^3 \cdot 10^{-5} = 10^{-2} \left(10^{-2} = \frac{1}{10 \cdot 10} = \frac{1}{100} = 0.01\right)$

2. $x^m / x^n = x^{m-n}$

   example: $10^6 / 10^2 = 10^4$

3. $x^0 = 1$ (Note that *any* quantity raised to the zero power equals 1.)

The process that is the inverse of raising to a power is called "extracting a root." For example, $\sqrt{9}$ means "what number multiplied by itself will yield 9?" Because $3 \times 3 = 9$, $\sqrt{9} = 3$. $\sqrt[5]{32}$ means "what number multiplied by itself 5 times will yield 32?" Because $2 \times 2 \times 2 \times 2 \times 2 = 32$, $\sqrt[5]{32} = 2$. You will not have to calculate roots of numbers because you will use your calculator to determine these values. However, you should know what operations your calculator is doing to get these values.

Example A.11

Find the value of $10^{-14}/10^{-6}$.

Solution

Using rule 2 above, we must subtract -6 from -14 to get the exponent for 10 in the answer. $-14 - (-6) = -14 + 6 = -8$. Therefore, $10^{-14}/10^{-6} = 10^{-14-(-6)} = 10^{-8}$.

Practice Exercise A.11

Find the value of $10^{-3} \cdot 10^{-5}$.

A.5 Scientific Notation and Rounding

I. Scientific Notation

Any quantity can be expressed using a power of ten by writing the quantity as 10, 100, 1000, etc. times a certain number. (In the textbook, this number is represented by the letter A. Therefore a number in scientific notation can be written generally as $A \times 10^{\text{exponent}}$.)

Consider the following examples:

$325 = 32.5 \times 10 = 3.25 \times 100 = 0.325 \times 1000$ or
$325 = 32.5 \times 10^1 = 3.25 \times 10^2 = 0.325 \times 10^3$

Expressing 325 in negative powers of ten, we obtain

$325 = 3250 \times 0.1 = 32500 \times 0.01 = 325000 \times 0.001$ or
$325 = 3250 \times 10^{-1} = 32500 \times 10^{-2} = 325000 \times 10^{-3}$

Because $10^0 = 1$ (rule 3) we can also express 325 as 325×10^0.

Scientific notation is the expression of a number as a power of ten multiplied by a number between 1 and 10. The number between 1 and 10 is called the *coefficient*. For example, when 325 is expressed as 3.25×10^2, 3.25 (the coefficient) is between 1 and 10, and it is multiplied by a power of 10 (10^2). To express a number in scientific notation, you move the decimal point to the position such that there is one nonzero digit to the left of the decimal point. If the original number has been made smaller by moving the decimal point, the sign of the exponent is positive and the value of the exponent is equal to the number of places the decimal point was moved. If the original number has been made larger, the sign of the exponent is negative. Consider the following example.

Example A.12

Express 0.0003821 in scientific notation.

Solution

$0.0003821 = 3.821 \times 10^{-4}$

Because 3.821 is larger than the original number (0.0003821), the sign of the exponent is negative, and because the decimal point was moved four places, the value of the exponent is -4.

Practice Exercise A.12

Express 56,873 in scientific notation.

Another way to view this is in terms of the direction in which the decimal point was moved. If the decimal point was moved to the left, the exponent is positive, and if the decimal point was moved to the right, the exponent is negative. We will use this method in the next example.

Example A.13

Express 5,288,000 in scientific notation.

Solution

$5,288,000 = 5.288 \times 10^6$

Because the decimal point in the original number was moved six places to the left to obtain the number 5.288, the sign of the exponent is positive, and the value of the exponent is 6. We also know that the sign of the exponent is positive because 5.288 is smaller than 5,288,000. Either way of thinking about the problem yields the same result. A mnemonic that may help you remember how to keep the signs straight is "Registered Nurses Love Patients," or RN, LP, which stands for right-negative; left-positive.

Practice Exercise A.13

Express 0.0000894 in scientific notation.

It is sometimes necessary to change a number from one power of ten to another power of ten. When changing to a new power of ten, we must consider the sizes of both the coefficient and the power of ten. In the number 3.2×10^5, the coefficient is 3.2 and the power of ten is 5. If we wish to express this as another power of 10 that is equal in value to 3.2×10^5, we must *increase* the power of ten if the coefficient *decreases*, and *decrease* the coefficient if the power of ten *increases*. This is shown in the illustration below.

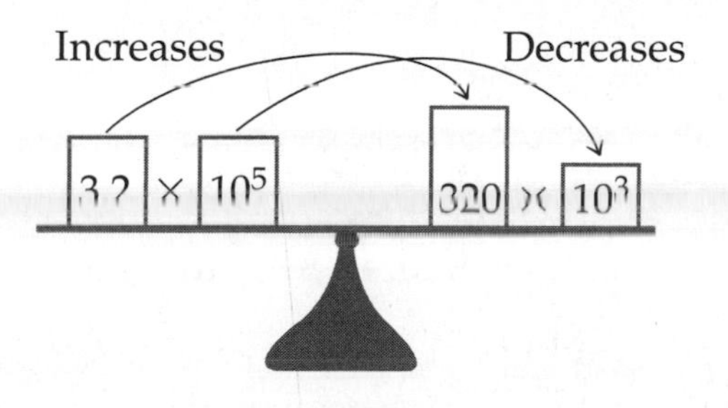

Example A.14

Fill in the blank: $6.79 \times 10^4 =$ ________ $\times 10^6$

Solution

Because the exponent has gotten larger by two places, the coefficient must get smaller by two places. We must change the 6.79 to 0.0679, which is the value that goes in the blank.

$$6.79 \times 10^4 = \underline{0.0679} \times 10^6$$

Practice Exercise A.14

Fill in the blanks:

(a) $5.87 \times 10^8 = 58.7 \times$ ______ (b) $473 \times 10^{-3} = 4.73 \times$ ______

Arithmetic Operations on Numbers Expressed as Powers of Ten

Although you will probably use your calculator to perform arithmetic operations on numbers expressed as powers of ten, you should know the rules that govern these processes. The rules are presented next.

A. Addition

1. Express the numbers as the same power of 10.
2. Add the coefficients.
3. Bring down the power of 10.
4. Express the answer in scientific notation. (In scientific notation the coefficient is between 1 and 10.)

Example A.15

Add $8.381 \times 10^{-4} + 5.83 \times 10^{-5}$

Solution

Make the exponents equal before adding the numbers. We can make both of the exponents 10^{-5} or we can make them both 10^{-4}. The result will be the same no matter which common exponent we choose. We'll make them both 10^{-4}. Therefore 5.83×10^{-5} must be changed, then we can add the two numbers.

$$\begin{array}{r} 8.381 \times 10^{-4} \\ +\ 0.583 \times 10^{-4} \\ \hline 8.964 \times 10^{-4} \end{array}$$

Practice Exercise A.15

Add $6.87 \times 10^8 + 3.56 \times 10^9$.

B. Subtraction

The rules are equivalent to the rules for addition; however, we subtract one number from the other after the exponents have been made the same. The specific rules are stated below:

1. Express the numbers as the same power of 10.
2. Perform the subtraction on the coefficients.
3. Bring down the power of 10.
4. Express the answer in scientific notation.

Example A.16

Subtract 4.72×10^2 from 5.32×10^3.

Solution

Again we must make the exponents equal. We'll make them both 10^2. Therefore 5.32×10^3 must be changed.

$$5.32 \times 10^3 = 53.2 \times 10^2$$

Now we can perform the subtraction.

$$\begin{array}{rl} & 53.20 \times 10^2 \\ - & \underline{\ \ 4.72 \times 10^2} \\ & 48.48 \times 10^2 \text{ or } 4.848 \times 10^3 \end{array}$$

Practice Exercise A.16

$0.0839 \times 10^8 - 6.29 \times 10^6$

C. Multiplication

To multiply two numbers expressed in scientific notation, multiply the coefficients and add the exponents on the powers of 10.

Example A.17

What is the product of 4×10^3 and 8×10^{-5}?

Solution

$4 \times 10^3 \times 8 \times 10^{-5} = 32 \times 10^{3+(-5)} = 32 \times 10^{-2} = 3.2 \times 10^{-1}$

Practice Exercise A.17

What is the product of 5.1×10^{-4} and 3.2×10^{-2}?

D. Division

To divide two numbers expressed in scientific notation, divide the coefficients and subtract the exponents on the powers of 10.

Example A.18

Divide 6.2×10^{-4} by 3.1×10^{-6}.

Solution

$$\frac{6.2 \times 10^{-4}}{3.1 \times 10^{-6}} = 2.0 \times 10^{-4-(-6)} = 2.0 \times 10^{2}$$

Practice Exercise A.18

Divide 4.0×10^{8} by 2.0×10^{3}.

II. Rounding Numbers

It is often necessary to round numbers after performing calculations. This is usually necessary when the calculated answer has more significant figures than are allowed. The rules for rounding numbers are presented below.

1. If the last digit to be kept is followed by a number less than 5, round down by dropping all digits to the right of it. For example, if 54.654873 is to be rounded to four digits, it will become 54.65. (The last digit to be kept is 5, and 4 is the number that follows it. Since 4 is less than 5, all digits to the right of the 5 are dropped.)
2. If the last digit to be kept is followed by a number greater than five, round up by adding 1 to the last digit to be kept and drop the rest. For example, if 0.8765 is to be rounded to two digits, it will become 0.88. (The last digit to be kept is 7. However, the 7 is followed by a 6, which is greater than 5. Therefore add 1 to the 7, and the number becomes 0.88.)
3. If the last digit to be kept is followed by a 5 and no other nonzero digits, round up by adding 1 if the last digit to be kept is odd, but round down if the last digit to be kept is even. For example, 5.875 rounded to three digits becomes 5.88, but 6.365 rounded to three digits becomes 6.36.
4. If the last digit to be kept is followed by a 5 and other nonzero digits, round up by adding 1. For example, 32.4521 rounded to three digits becomes 32.5.

A.6 Logarithms and Inverse Logarithms

A. Logarithms

The common logarithm, or base ten logarithm (abbreviated "log"), of a number is the power to which 10 must be raised in order to obtain the number. For example, log 2 = 0.30 means that $10^{0.30} = 2$. The log of $10^{-5} = -5$, and the log of $10^8 = 8$. The common logarithm of a number can be obtained by pressing the LOG button on your calculator.

It is important to remember that the logarithm of any number less than one is negative, and the logarithm of any number greater than 1 is a positive number. The log of 1 = 0, since $10^0 = 1$.

Example A.19

What is the logarithm of 10^{-3}?

Solution

Since 10 must be raised to the power of −3 to obtain 10^{-3}, the log of $10^{-3} = -3$.

Practice Exercise A.19

What is the logarithm of 10^4?

B. Inverse Logarithms (Antilogarithms)

Finding a number when its logarithm is given is known as taking the inverse log, or antilog, of a number. This operation is done on the calculator by pressing the 10^x button.

Example A.20

What is the inverse logarithm of 3?

Solution

We can readily see that 10^3 is the inverse log of 3. (10 raised to the third power is 10^3.) Thus the answer is 1000.

Practice Exercise A.20

What is the inverse logarithm of −9?

A.7 Ratios and Proportions

A. Ratios

A ratio is equivalent to comparing two numbers by division. Ratios can be expressed in three different ways. The ratio of a to b can be expressed as: a to b, $a{:}b$, or a/b. Since a ratio can be thought of as a fraction (a/b), it can also be reduced to lowest terms. Note that the ratio of b to a will be expressed as b to a, $b{:}a$, or b/a. The ratio of b/a is simply the inverse of the ratio of a/b. (Recall that the inverse of a fraction is a fraction in which the numerator and denominator have been reversed; the fraction has been turned "upside down.")

The ratio of the number of months in a year to the number of hours in a day can be expressed as 12 to 24, 12:24, or 12/24. If we reduce the fraction 12/24 to lowest terms, we see that the ratio can also be expressed as 1 to 2, 1:2, or 1/2. We could have expressed the ratio as any fraction that is equivalent to 12/24, such as 6/12 or 3/6.

If we had chosen to represent this ratio as the ratio of the number of hours in a day to the number of months in a year, the ratio would have been expressed as 24 to 12, 24:12, or 24/12. Reduced to lowest terms the ratio would have been 2 to 1, 2:1, or 2/1.

Example A.21

If the number of women in a chemistry class is 48, and the number of men is 64, what is the ratio of women to men in the class? What is the ratio of men to women?

Solution

The ratio of women to men is 48:64 (48/64), or 3:4 (3/4).

The ratio of men to women is 64:48 (64/48), or 4:3 (4/3).

Practice Exercise A.21

(a) If the mass of carbon is 12, and the mass of cobalt is 60, what is the ratio of the mass of carbon to the mass of cobalt?

(b) What is the ratio of the mass of cobalt to the mass of carbon?

B. Proportions

A proportion is a statement that two ratios are equal. For example:

$$\frac{52}{4} = \frac{26}{2}$$

Proportions can be read out loud as:

52 is to 4 as 26 is to 2. Or they can be written as 52:4 = 26:2.

Proportions are very useful in problem-solving, and are used often in chemistry. In most problems involving proportions, we will need to find one of the values in a proportion. For example, we may know that it takes 16 units of oxygen to make 32 units of water, but we want to know how many units of water can be made from 4 units of oxygen. We can set this up as a proportion. 16 units of oxygen is to 32 units of water as 4 units of oxygen is to ? units of water. 16 units of oxygen:32 units of water as 4 units of oxygen: x units of water.

We can write this in fractional form as:

$$\frac{16 \text{ units of oxygen}}{32 \text{ units of water}} = \frac{4 \text{ units of oxygen}}{x \text{ units of water}}$$

To solve for x, we can cross-multiply the numbers to obtain an algebraic equation.

$$16 \cdot x = 4 \cdot 32$$
$$16x = 128$$

We divide both sides by 16 to find that $x = 8$. Therefore, 4 units of oxygen will make 8 units of water.

It is important to include the units when setting up proportions. A common mistake students make is to reverse the numbers when setting up the proportion. This will lead to an incorrect answer. The same units *must* be across from each other when setting up the fractional form of the proportion.

Example A.22

If a chemical reaction produces 8 grams of product in 2 hours, how many grams of product will be produced in 24 hours, assuming that the rate of production remains the same?

Solution

We can set up a proportion as follows:

$$\frac{8 \text{ grams}}{2 \text{ hours}} = \frac{x \text{ grams}}{24 \text{ hours}}$$

Cross-multiplying, we obtain the equation $8 \cdot 24 = 2x$, or $192 = 2x$. Solving for x, we get $x = 96$. Therefore, the reaction will produce 96 grams of product in 24 hours.

Practice Exercise A.22

If 46 grams of sodium are needed to make 58 grams of sodium chloride, how many grams of sodium are needed to make 290 grams of sodium chloride?

A.8 Length, Area, Volume

In chemistry, you will often work with length, area, and volume measurements. It is important that you understand the relationships among these properties. Length is measured in centimeters, meters, feet, yards, kilometers, etc. These are abbreviated cm, m, ft, yd, and km, respectively. Note that there are one hundred centimers in a meter, and a thousand meters in a kilometer. Area is expressed as a length unit multiplied by a length unit. The units of area can be expressed as the square of a length unit. For example, areas are expressed as square meters, square feet, square yards, square kilometers, etc. These are written as m^2, ft^2, yd^2, or km^2. Volume can be represented as a length unit cubed. For example, volumes are expressed as cubic centimeters, cubic meters, cubic feet, cubic yards, or cubic kilometers. These are written as cm^3, m^3, ft^3, yd^3, and km^3, respectively. Volume can also be expressed as liters, gallons, etc. There are one thousand cubic centimeters (cm^3) in a liter. Note that 1 cubic centimeter is equal to 1 milliliter, or 1 cm^3 = to 1 mL.

We should know the formulas for calculating the areas and volumes of certain geometric shapes. The most important formulas are given below.

Area

For squares: area = length squared (l^2)

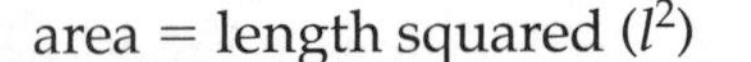

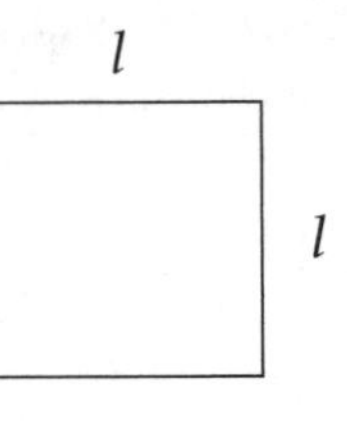

For rectangles: area = length · width

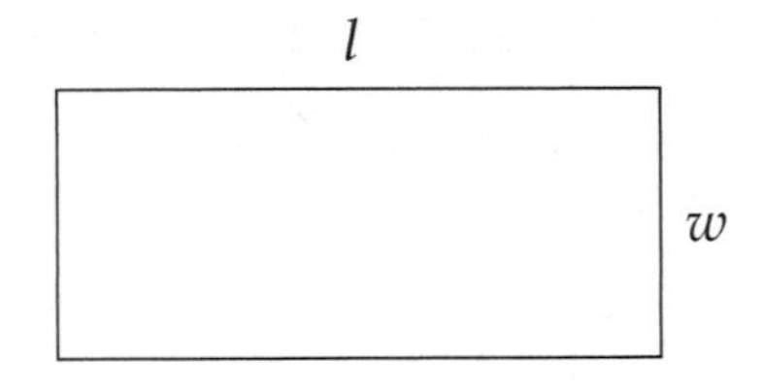

For circles: area = $\pi \cdot r^2$, where r represents the radius of the circle

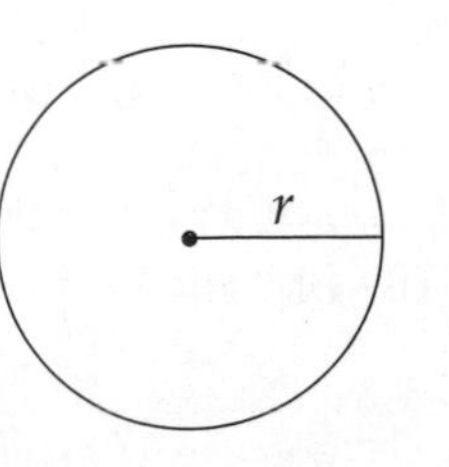

Volume

For cubes: volume = length cubed (l^3)

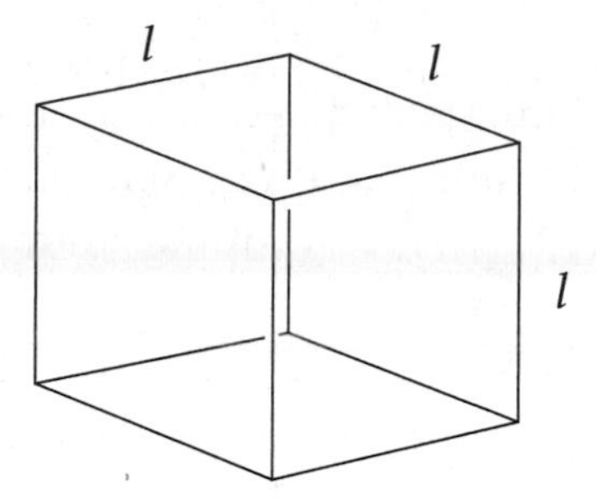

For rectangular solids: volume = length · width · height

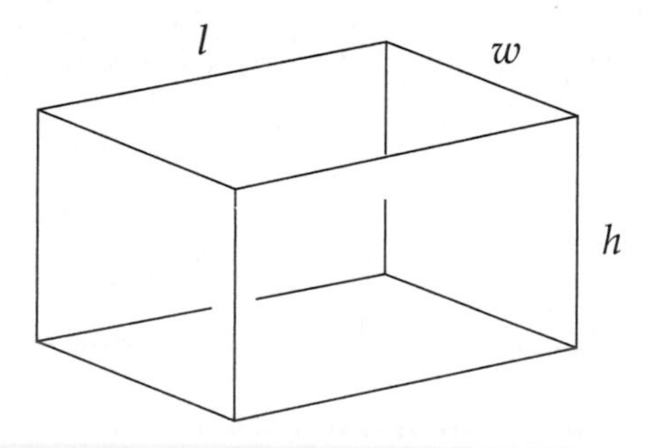

Example A.23

A rectangular solid measures 2 feet by 3 feet by 4 feet. What is the volume of the solid?

Solution

The volume of a rectangular solid is length · width · height.

$$\text{Volume} = 2 \text{ ft} \cdot 3 \text{ ft} \cdot 4 \text{ ft} = 24 \text{ ft}^3$$

Practice Exercise A.23

A cube measures 5 meters on each side. What is the volume of the cube?

A.9 Averages and Weighted Averages

The average of a set of numbers is the sum of the values divided by the number of values in the set.

$$\text{Average value} = \frac{\text{sum of all values}}{\text{\# of values}}$$

For example, the average of 5, 10, and 12 is 9.

$$\frac{5+10+12}{3}=9$$

Sometimes, however, the values in a set of numbers do not contribute equally to the set. For example, in determining your grade point average (GPA) in school, if you earned an A in one three-credit course, and a C in another three-credit course, your average would be exactly a B. However, if you earned an A in a two-credit course, and a C in a four-credit course, your average would be lower than a B. The A would count half as much toward your GPA as the C, because the course in which you earned the A is worth half as many credit hours. In this case, we calculate the *weighted* average.

$$\text{Weighted average} = \frac{(\text{factor for value 1})(\text{value 1}) + (\text{factor for value 2})(\text{value 2}) + \ldots}{\text{sum of all values}}$$

Let's look at the two cases in detail.

Case 1: Determining the average of two grades when the two courses are equal in credit hours. The points assigned to each grade are as follows:

A = 4 points B = 3 points C = 2 points D = 1 point

$$\text{Average value} = \frac{\text{sum of all values}}{\text{\# of values}}$$

$$\text{Average of one A and one C} = \frac{4+2}{2} = \frac{6}{2} = 3$$

This result is exactly what we expect. An A and a C will yield a B average.

Case 2: Determining the average of two grades when the two courses are not equal in credit hours. You earned an A in a three-credit course, and a C in a four-credit course. In this example the factor for the values is the number of credit hours, and the value is the number of points associated with the particular grade.

$$\text{Weighted average} = \frac{(\text{factor for value 1})(\text{value 1}) + (\text{factor for value 2})(\text{value 2}) + \ldots}{\text{sum of the weighting factors}}$$

$$\text{Weighted average} = \frac{(3)(4)+(4)(2)}{3+4} = \frac{20}{7} = 2.85$$

You will use weighted averages when you calculate the average atomic weights of elements.

Example A.24

Chlorine has two naturally occurring isotopes. Isotope A has a mass of 34.97, and represents 75.8% of naturally occurring chlorine. Isotope B has a mass of 36.97, and represents 24.2% of naturally occurring chlorine. What is the average atomic mass of chlorine?

Solution

The weighting factors are the percentages, and the values are the masses. Remember to change the percentages to decimals when performing the calculations.

$$\text{Weighted average} = \frac{(\text{factor for value 1})(\text{value 1}) + (\text{factor for value 2})(\text{value 2}) + \ldots}{\text{sum of the weighting factors}}$$

$$\text{Average atomic mass} = \frac{(0.758)(34.97) + (0.242)(36.97)}{0.758 + 0.242} = \frac{35.45}{1.000} = 35.45$$

Practice Exercise A.24

Boron has two naturally occurring isotopes. Isotope 1 has a mass of 10.01, and represents 19.8% of naturally occurring boron. Isotope 2 has a mass of 11.01, and represents 80.2% of naturally occurring boron. What is the average atomic mass of boron?

A.10 Graphs

Graphs are used in chemistry to show relationships between two variables. They indicate how one variable changes when the other is changed. Variables can be *directly* related to each other (directly proportional) or *inversely* related (also called inversely proportional). When variables are directly proportional, as one increases, so does the other, as in the figure below:

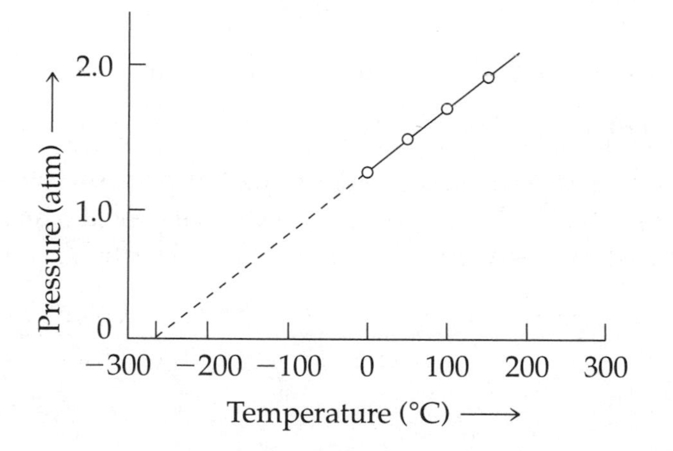

This graph shows that as the temperature increases, so does the pressure. The axis that shows the temperature values is the x-axis, and the axis that shows the pressure is the y-axis. Notice that the differences between the temperature values shown on the x-axis and the pressure values shown on the y-axis are constant—100 degrees Celsius for each temperature interval on the x-axis, and 1.0 atm for each pressure interval on the y-axis. This graph shows a *direct* relationship between the variables because as the temperature increases, so does the pressure. At 0°C the pressure is approximately 1.1 atm, and at 200°C the pressure is approximately 2.0 atm.

Example A.25

What is the approximate pressure when the temperature is 100°C?

Solution

To determine the pressure at a temperature of 100°C, we draw a vertical line from 100° on the x-axis up to the graph. We then draw a horizontal line from this point on the graph to the y-axis. The point at which the horizontal line intersects the y-axis is the value of the pressure when the temperature is 100°C. The value of the pressure is approximately 1.8 atm. This process is shown in the figure below.

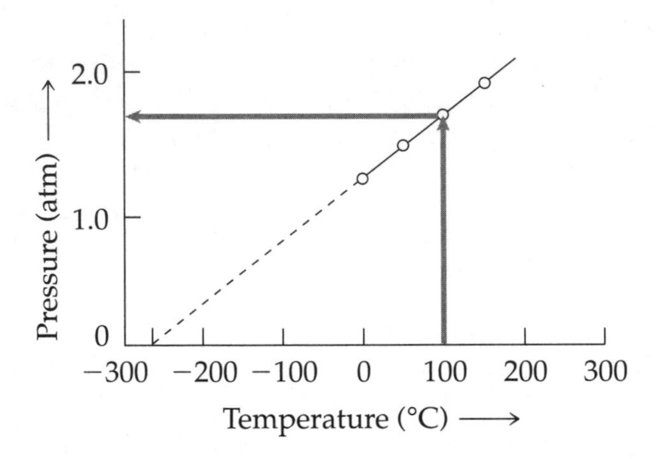

Practice Exercise A.25

What is the approximate temperature when the pressure is 1.5 atm?

When variables are inversely related to each other, as one variable increases, the other variable decreases. The graph below shows the inverse relationship between the pressure and the volume of a gas. Notice that as the pressure increases, the volume decreases, and vice versa.

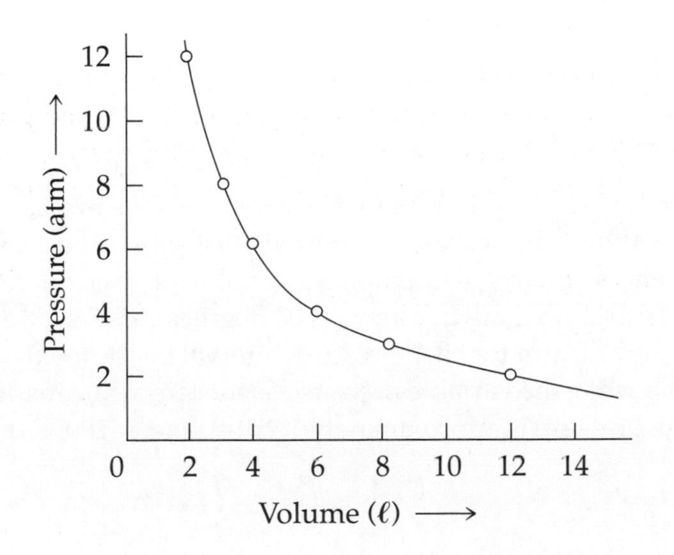

Example A.26

What is the volume occupied by a gas that is under a pressure of 4 atm?

Solution

We draw a horizontal line from the value of 4 atm on the y-axis over to the curve. We then draw a vertical line from this point on the curve to the x-axis. The point at which the vertical line intersects the x-axis is the value of the volume when the pressure is 4 atm. The volume is 6 liters. This process is shown in the figure below.

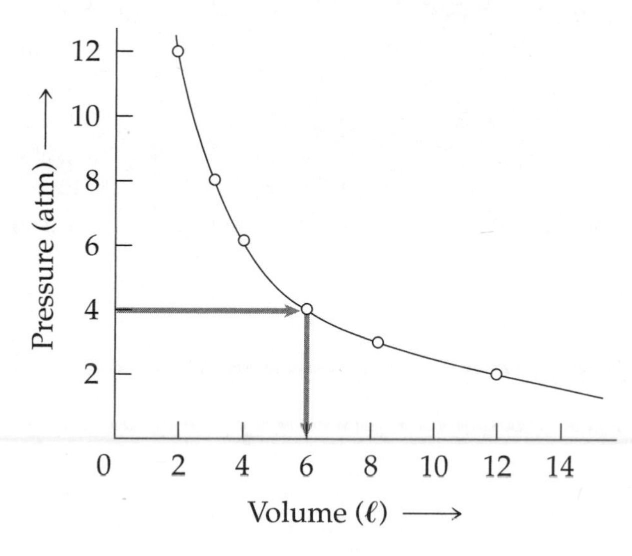

Practice Exercise A.26

What is the pressure of a gas that is occupying a volume of 4 liters?

A.11 Unit Analysis and Conversion Factors

An important part of problem-solving in chemistry is calculating new quantities from information that is given. Calculating the new quantities often involves ratios and proportions. For example, we may be told that one mole of carbon has a mass of 12 grams, and we may be asked to find the number of moles in 54 grams of carbon. In Section 1.7 we learned a method for doing this type of problem.

$$\frac{\text{1 mole of carbon}}{\text{12 grams of carbon}} = \frac{x \text{ moles of carbon}}{\text{54 grams of carbon}}$$

Cross-multiplying, we obtain $54 = 12x$. Solving for x, we obtain $x = 4.5$. Therefore, 54 grams of carbon = 4.5 moles of carbon.

Another way of solving this problem involves using the known ratio (or its inverse) as a *conversion factor*. This conversion factor is multiplied by the value given in the problem in such a way as to cancel all units except the unit in which the answer will be expressed. It is important to remember that identical units will cancel only when one is in the numerator and the other is in the denominator. Using this method in the problem above, we would have:

$$54 \text{ grams of carbon} \times \frac{1 \text{ mole of carbon}}{12 \text{ grams of carbon}} = 4.5 \text{ moles of carbon}$$

This method is very helpful if you are not sure how to start a problem. Just start with the value given, and multiply it by conversion factors (ratios) until you get the desired unit.

Example A.27

There are 16 weeks in an average semester in college. How many hours is this?

Solution

We must find the number of hours in 16 weeks. The ratios (or their inverse) we will need to use are:

$$\frac{7 \text{ days}}{1 \text{ week}} \qquad \frac{24 \text{ hours}}{1 \text{ day}}$$

We begin with the given value, and use the appropriate ratios.

$$16 \text{ weeks} \times \frac{7 \text{ days}}{1 \text{ week}} = \frac{24 \text{ hours}}{1 \text{ day}} = 2688 \text{ hours}$$

Practice Exercise A.27

The distance between your residence hall and your chemistry class is 378 yards. What is this distance in centimeters (cm)? The ratios you will need are:

$$\frac{1 \text{ yard}}{3 \text{ feet}} \qquad \frac{1 \text{ foot}}{12 \text{ inches}} \qquad \frac{1 \text{ inch}}{2.54 \text{ cm}}$$

A.12 Probability

Although you will not do any calculations involving probability, you must understand the meaning of the concept. The probability that an event will occur is often expressed as a percentage. For example, "the probability of rain is 95%" means that there is a great chance that rain will occur. Probability expressed as a percent states the number of times that an event is likely to occur if given 100 chances to occur. Thus, a 95% probability indicates that rain is likely to occur 95 times in 100 chances. Events that have less than a 50% probability will likely occur less than half of the times in which they are given a chance to occur.

A.13 Geometric Figures

Geometric figures, and the angles associated with them, are an important part of chemistry, especially when discussing the shapes of molecules and the bond angles in molecules. Some basic principles are presented below.

A. Angles

1. The angle of a circle is 360° (see the figure below).

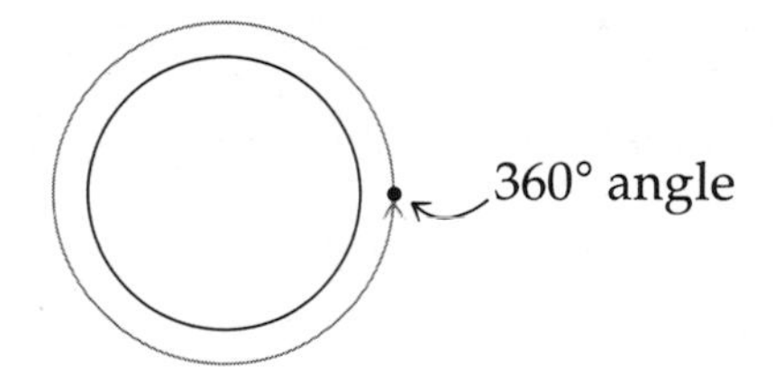

2. The angle of a straight line is 180° (see the figure below).

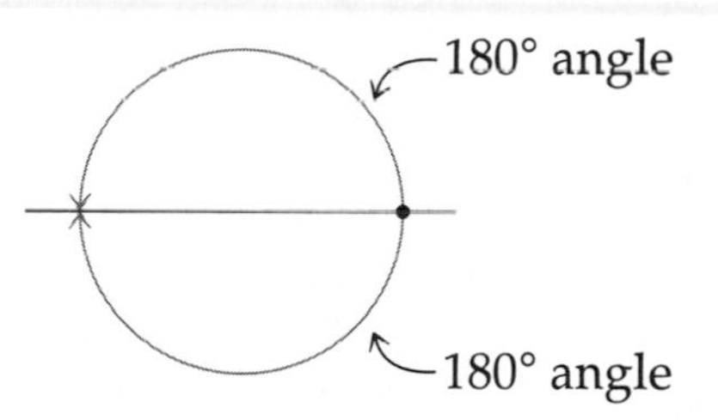

3. The angles at the intersection of three lines equally spaced around a circle are 120° (see the figure below).

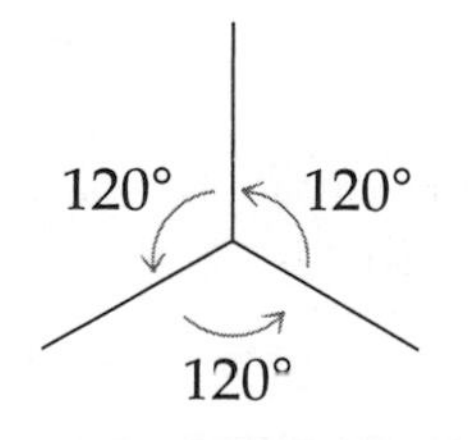

4. The angles at the intersection of two lines perpendicular and crossing each other are 90° (see the figure below).

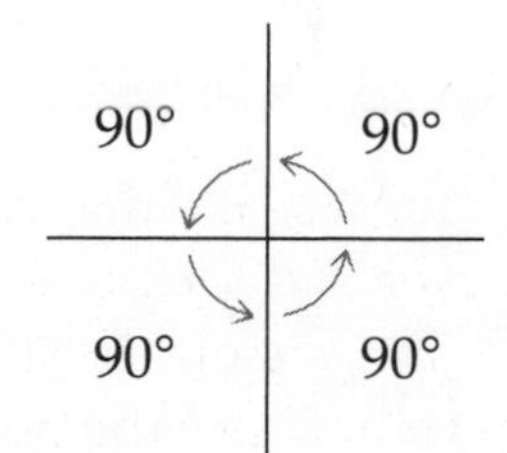

Knowing the values of these angles will be quite handy when describing the arrangement of electrons in molecules.

Example A.28

The BF_3 molecule, shown below, has a shape that is called trigonal planar. What are the bond angles in the molecule?

```
F     F
 \   /
   B
   |
   F
```

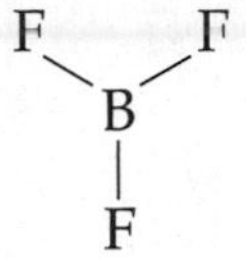

Solution

Since the angles are formed by the intersection of three lines that come together at the center, the bond angles are 120°.

Practice Exercise A.28

The $BeCl_2$ molecule is linear, as shown below. What are the bond angles in the molecule?

Cl—Be—Cl

Solutions to Practice Problems and Quizzes

Chapter 1
What is Chemistry?

Practice Problems

1.1 Heterogeneous mixture
1.2 Elemental substance
1.3 Compound
1.4 Homogeneous mixture
1.5 Chemical
1.6 Physical
1.7 Physical
1.8 Chemical

Quiz

1. (a) Heterogeneous mixture
   (b) Homogeneous mixture
   (c) Heterogeneous mixture
   (d) Elemental substance
2. (a) Physical
   (b) Chemical
   (c) Physical
3. New substance
4. (a) Compound
   (b) Elemental substance
   (c) Compound
   (d) Elemental substance
5. (a) Physical
   (b) Chemical
   (c) Chemical
   (d) Physical
6. Homogeneous mixtures are mixtures whose composition is constant throughout. Heterogeneous mixtures have variable composition throughout the mixture.
7. The milk would begin to develop a strong sour odor. It would also appear to be very clumpy. These changes indicate that a new substance has been produced; thus a chemical transformation must have taken place.
8. (Answers may vary.)
   (a) Chlorine gas
   (b) Carbon dioxide
   (c) The air we breathe
   (d) Chocolate-chip ice cream
9. False. Fizzing seltzer water contains not only water but also carbon dioxide, which produces the fizz.
10. Chemical

Chapter 2
The Numerical Side of Chemistry

Practice Problems

2.1 2
2.2 8
2.3 5
2.4 4
2.5 4
2.6 6
2.7 4
2.8 1
2.9 3
2.10 0.006 74

2.11 6740
2.12 8,342,100,000
2.13 0.000 035 8
2.14 8.3×10^7
2.15 7.8200×10^{-4}
2.16 5.03×10^2
2.17 4.67×10^{-1}
2.18 2.9×10^1
2.19 6×10^{-4}
2.20 5.86×10^{-2}
2.21 3.9×10^{22}
2.22 937
2.23 5.175×10^{-2}
2.24 6.3×10^1
2.25 13.375
2.26 1.9
2.27 34
2.28 36.3
2.29 357
2.30 3.1
2.31 7.7×10^{-2}
2.32 31.3
2.33 0.0127 L of iron
2.34 1.2 mi/min
2.35 0.000 536 or 5.36×10^{-4}
2.36 250,000 cm^2
2.37 \$0.17/in.^{2}
2.38 4.4×10^4 J
2.39 136°C

Quiz

1. (a) 3
(b) 5
(c) 3
(d) 2
(e) 7

2. (a) 2 zeros
(b) 4 zeros
(c) 3 zeros
(d) 2 zeros
(e) 0 zeros

3. (a) 0.008 59
(b) 276
(c) 0.0276
(d) 7,200,000,000

4. (a) 8.304×10^{-10}
(b) 9.5×10^6
(c) 1.3×10^{-2}
(d) 5.83×10^1
(e) 5.83×10^{-1}

5. (a) 51.8
(b) 5.32
(c) 244.9
(d) 1583

6. (a) 5.0
(b) 2.0
(c) 3.0×10^4
(d) 2.2×10^2

7. (a) 46
(b) 2.2×10^5
(c) 5.6
(d) 6

8. (a) 0.0583 L
(b) 4.63×10^{-5} m
(c) 8.39×10^5 g
(d) 594 cm^2

9. 14.0 mL

10. 1.02×10^3 g

11. 305 in.^{3}

12. \$0.21/in.^{2}

13. 2.3×10^4 J

14. 2.7×10^{-3} g

15. 120 km/h

16. 156,000. indicates that there are six significant figures. 156,000 has only three significant figures. 156,000. assumes a more accurate measuring device.

17. 99°F

18. (a) Less than
(b) Equal to
(c) Less than
(d) Equal to
(e) Less than

19. 19.3 g/mL

20. 356.3 K

Chapter 3
The Evolution of Atomic Theory

Practice Problems

3.1 15.79% C; 84.21% S

3.2 41.03% O; 58.97% Na

3.3 52.9% Al; 47.1% O

3.4 80.1%

3.5 39.948 amu

3.6 I < Br < Cl < F

3.7 I > Br > Cl > F

3.8 Sn, because it is lower on the periodic table.

Quiz

1. 3.25% H; 19.37% C; 77.38% O
2. 107.87 amu
3. Po < Pb < Tl < Cs
4. ^{39}K (38.9637 amu) will be present in the greatest amount because its atomic mass is the closest to the weighted average atomic mass of 39.098 amu.
5. Be < C < O < F < Ne
6. Tl
7. 95.71%
8. Cs > Rb > K > Na > Li
9. Xe
10. Ne > F > O > C > Be

Cumulative Quiz for Chapters 1, 2, and 3

1. (a) Heterogeneous mixture
   (b) Homogeneous mixture
   (c) Compound
   (d) Elemental substance
   (e) Homogeneous mixture
2. (a) Chemical
   (b) Physical
   (c) Chemical
3. False. A substance is only a compound if more than one element is present. Because S is the only element present, S_8 is not a compound.
4. element, compound
5. (a) 2 (b) 5 (c) 3 (d) 2
6. (a) 2 (b) 4 (c) 0 (d) 0
7. (a) 7,450,000 (b) 81,200
   (c) 0.341 (d) 730,000,000
8. (a) 7.842×10^{-4} (b) 8.6×10^{7}
   (c) 1.23×10^{-1} (d) 5.83×10^{1}
9. (a) 78.3 (b) 60.0 (c) 0.0054
10. (a) 23 (b) 4.6×10^{6} (c) 5.6
11. (a) 927,000 mL (b) 56.4 cm
    (c) 0.0563 m^2
12. 634 mL
13. 483 g
14. 1.05×10^{5} J
15. 639 g
16. 50 km/h
17. 212°F
18. C: 40.0%; H: 6.71%; O: 53.28%
19. 63.55 amu
20. Sn < Sb < Te < I < Xe

Chapter 4
The Modern Model of the Atom

Practice Problems

4.1 5.68×10^{-19} J

4.2 4.86×10^{-7} m or 486 nm

4.3 4.09×10^{-19} J

4.4 3 eV

4.5 1 eV

4.6 13.2 eV

4.7 20 electrons: $1s^22s^22p^63s^23p^64s^2$

4.8 34 electrons: $1s^22s^22p^63s^23p^64s^23d^{10}4p^4$

4.9 82 electrons:
$1s^22s^22p^63s^23p^64s^23d^{10}4p^65s^24d^{10}5p^66s^24f^{14}5d^{10}6p^2$

4.10 $[Ne]3s^2$

4.11 $[Xe]6s^24f^{14}5d^{10}6p^3$

4.12 $[Ar]4s^23d^3$

Quiz

1. 1.1 eV
2. 4.97×10^{-19} J
3. 7 electrons: $1s^22s^22p^63s^23p^64s^1$

4. Krypton: [Kr]
5. 12.8 eV
6. 2.7×10^{-19} J
7. 15 electrons: $1s^2 2s^2 2p^6 3s^2 3p^3$
8. The scaled energies are equivalent. The amount of energy released and absorbed would be 0.1 eV.
9. 48 electrons: [Kr] $5s^2 4d^{10}$
10. 5.68×10^{-8} m or 56.8 nm

Chapter 5
Chemical Bonding and Nomenclature

Practice Problems

5.1 H:Ö:Ö:H

5.2 H:C::C:H (each C also bonded to H above)

5.3 [:Ö:Cl:Ö: / :Ö:]$^-$

5.4 Barium chloride
5.5 Cesium nitride
5.6 Potassium iodide
5.7 Cobalt(II) iodide
5.8 Cuprous fluoride
5.9 Mercury(II) bromide
5.10 Diboron trioxide
5.11 Nitrogen trifluoride
5.12 Xenon tetrafluoride
5.13 Titanium(III) fluoride
5.14 Nickel(II) oxide
5.15 Tin(IV) chloride
5.16 Chlorous acid
5.17 Perbromic acid
5.18 Hydrofluoric acid

Quiz

1. (a) :Cl:P:Cl: / :Cl:

   (b) :Ö::S:Ö:

2. (a) [H:N:H with H above and below N]$^+$

   (b) [:Ö::N:Ö:]$^-$

3. (a) Calcium phosphide
   (b) Iron(III) oxide
4. (a) Sulfuric acid (*Note:* this was given in a table in the text.)
   (b) Nitrous acid (*Note:* this was given in a table in the text.)
5. +2, +1
6. (a) Magnesium nitride
   (b) Nitrous acid
   (c) Hypochlorous acid
7. (a) 32 valence electrons
   (b) 32 valence electrons
   (c) 8 valence electrons
   (d) 18 valence electrons
8. (d) Hydrogen
9. (b) $CoCl_2$, (d) $HgCl_2$, and (e) $CrCl_2$
10. Nitrogen and oxygen form more than one binary molecular compound. If there are no prefixes in the name to indicate the number of nitrogens and oxygens in the formula, it is impossible to tell whether the compound's formula is NO, N_2O, NO_2, N_2O_5, or something else.

Chapter 6
The Shape of Molecules

Practice Problems

6.1 Bent; 105° (two lone pairs)
6.2 Trigonal planar; 120°
6.3 Tetrahedral; 109.5°
6.4 Linear; 180°
6.5 PI_3 is a polar molecule; the net dipole moment is in the direction of the iodines.
6.6 ONCl is a polar molecule. The shape is bent with the net dipole in the direction of the oxygen atom.

6.7 No net dipole exists. BCl_3 is a trigonal planar molecule; the bond dipole moments cancel each other.

Quiz

1. Trigonal planar; 120°

```
 :O:
  ::  ..
H:C:Cl:
     ..
```

2. (a) Linear (b) Trigonal planar
   (c) Tetrahedral
3. Pyramidal
4. (a) Nonpolar
   (b) Polar; the molecular dipole moment is in the direction of the F.

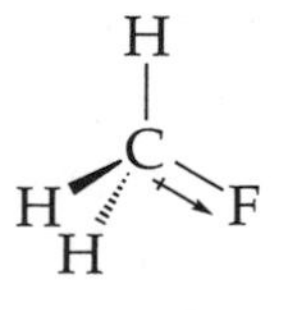

   (c) Polar; the molecular dipole moment is in between the F's.

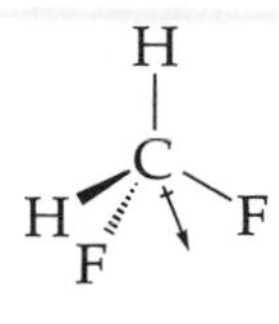

   (d) Polar; the molecular dipole moment is in the direction of the F's.

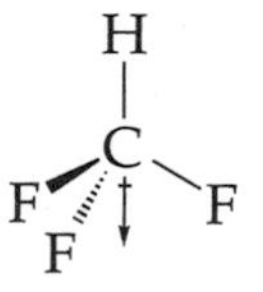

   (e) Nonpolar
5. C–H < C–Br < C–Cl < C–O < C–F
6. False. When one or more lone pairs are present, the molecule will be pyramidal or bent.
7. Linear
8. BF_3, SO_3
9. Tetrahedral, bent
10. For the molecular shape with bond angles of ~105°:
    (a) Four electron groups
    (b) Two bonding pairs
    (c) Two lone pairs

    For the molecular shape with bond angles of ~118°:
    (a) three electron groups
    (b) two bonding pairs
    (c) one lone pair

Cumulative Quiz for Chapters 4, 5, and 6

1. 2.6 eV
2. 8 electrons
3. 4.18×10^{-19} J
4. 38 electrons: $1s^2 2s^2 2p^6 3s^2 3p^6 4s^2 3d^{10} 4p^6 5s^2$
5. 57 nm
6. (a)

```
 ..   ..  ..
:Br : N : Br:
 ..   ..  ..
     :Br:
      ..
```

   (b)

```
 ..     ..
:F : B : F:
 ..  ..  ..
    :F:
     ..
```

7. (a)

```
 [ ..      ..  ]2−
 [:O : C :: O: ]
 [ ..  ..      ]
 [    :O:      ]
 [     ..      ]
```

   (b)

```
 [    H    ]+
 [    ..   ]
 [ H : N : H ]
 [    ..   ]
 [    H    ]
```

8. (a) Magnesium nitride
   (b) Cobalt(II) sulfide
   (c) Chloric acid
   (d) Nitrous acid
9. (a) 26 electrons
   (b) 24 electrons
   (c) 20 electrons
10. Hydrogen and boron
11.

```
  ..
 :Br:
  ..
H:C:H
  ..
  H
```

    Molecular shape: tetrahedral
    Bond angle: 109.5°
12.

```
 [ ..   ..  .. ]2−
 [:O : S : O: ]
 [ ..   ..  .. ]
 [     :O:     ]
 [      ..     ]
```

    Molecular shape: pyramidal
    Bond angle: 107°

13. (a) Nonpolar

(b) Polar; the net dipole moment is between the Cl's.

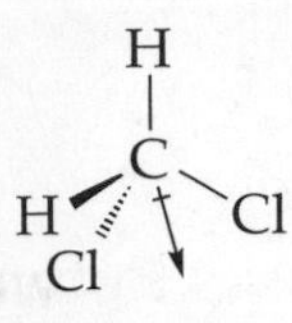

(c) Polar; the net dipole moment is between the H's.

(d) Nonpolar

14. Bent
15. BI_3
16. Trigonal planar, bent
17. Pyramidal: four bonding groups and one lone pair

    Trigonal planar: three bonding groups and no lone pairs
18. H–H < H–Br < H–Cl < H–O < H–F
19. Nonpolar
20. (a) 120° (b) 107° (c) 109.5°

Chapter 7
Chemical Reactions

Practice Problems

7.1 $NCl_3 + 3\,H_2O \longrightarrow NH_3 + 3\,HOCl$

7.2 $2\,Al_2O_3 \longrightarrow 4\,Al + 3\,O_2$

7.3 $PCl_5 + 4\,H_2O \longrightarrow H_3PO_4 + 5\,HCl$

7.4 Balanced equation:

$2\,NaOH(aq) + MgCl_2(aq) \longrightarrow Mg(OH)_2(s) + 2\,NaCl(aq)$

Ionic equation:

$2\,Na^+(aq) + 2\,OH^-(aq) + Mg^{2+}(aq) + 2\,Cl^-(aq) \longrightarrow Mg(OH)_2(s) + 2\,Na^+(aq) + 2\,Cl^-(aq)$

Net ionic equation:

$2\,OH^-(aq) + Mg^{2+}(aq) \longrightarrow Mg(OH)_2(s)$

7.5 Balanced equation:

$K_3PO_4(aq) + Al(NO_3)_3(aq) \longrightarrow AlPO_4(s) + 3\,KNO_3(aq)$

Ionic equation:

$3\,K^+(aq) + PO_4^{3-}(aq) + Al^{3+}(aq) + 3\,NO_3^-(aq) \longrightarrow AlPO_4(s) + 3\,K^+(aq) + 3\,NO_3^-(aq)$

Net ionic equation:

$PO_4^{3-}(aq) + Al^{3+}(aq) \longrightarrow AlPO_4(s)$

7.6 Molecular neutralization reaction equation:

$HBr(aq) + KOH(aq) \longrightarrow KBr(aq) + H_2O(l)$

Ionic neutralization reaction equation:

$H^+(aq) + Br^-(aq) + K^+(aq) + OH^-(aq) \longrightarrow K^+(aq) + Br^-(aq) + H_2O(l)$

Net ionic equation:

$H^+(aq) + OH^-(aq) \longrightarrow H_2O(l)$

7.7 Molecular neutralization reaction equation:

$2\,HNO_3(aq) + Ba(OH)_2(aq) \longrightarrow Ba(NO_3)_2(aq) + 2\,H_2O(l)$

Ionic neutralization reaction equation:

$2\,H^+(aq) + 2\,NO_3^-(aq) + Ba^{2+}(aq) + 2\,OH^-(aq) \longrightarrow Ba^{2+}(aq) + 2\,NO_3^-(aq) + 2\,H_2O(l)$

Net ionic equation:

$2\,H^+(aq) + 2\,OH^-(aq) \longrightarrow 2\,H_2O(l)$, which reduces to $H^+(aq) + OH^-(aq) \longrightarrow H_2O(l)$

Quiz

1. $C_{10}H_{20} + 15\,O_2 \longrightarrow 10\,CO_2 + 10\,H_2O$
2. $4\,PH_3 + 9\,O_2 \longrightarrow P_4O_{12} + 6\,H_2O$; the coefficient is 4.
3. $MgCO_3$ and $CaSO_4$ are insoluble in water.
4. $CaCO_3$
5. Li^+, Cl^-, Ba^{2+}, NO_3^-
6. Molecular equation:

   $Na_2SO_4(aq) + Pb(NO_3)_2(aq) \longrightarrow PbSO_4(s) + 2\,NaNO_3(aq)$

   Ionic equation:

   $2\,Na^+(aq) + SO_4^{2-}(aq) + Pb^{2+}(aq) + 2\,NO_3^-(aq) \longrightarrow PbSO_4(s) + 2\,Na^+(aq) + 2\,NO_3^-(aq)$

   Net ionic equation:

   $Pb^{2+}(aq) + SO_4^{2-}(aq) \longrightarrow PbSO_4(s)$
7. Salt, water
8. Molecular equation:

   $2\,NH_4I(aq) + Pb(NO_3)_2(aq) \longrightarrow PbI_2(s) + 2\,NH_4NO_3(aq)$

Ionic equation:

$2\ NH_4^+(aq) + 2\ I^-(aq) + Pb^{2+}(aq) + 2\ NO_3^-(aq) \longrightarrow PbI_2(s) + 2\ NH_4^+(aq) + 2\ NO_3^-(aq)$

Net ionic equation:

$Pb^{2+}(aq) + 2\ I^-(aq) \longrightarrow PbI_2(s)$

9. $CaSO_4$ is insoluble in water.
10. Neutralization
11. Coefficient, subscripts

Chapter 8
Stoichiometry and the Mole

Practice Problems

8.1 70.56 g NH_3
8.2 10.20 mol NaOH
8.3 225.2 g H_2O
8.4 61.8 g CO_2
8.5 C_3H_3O
8.6 CH_2
8.7 $C_6H_6O_2$
8.8 $C_{23}H_{46}$
8.9 Sn: 61%; F: 39%
8.10 Mg: 16.4%; N: 18.9%; O: 64.7%
8.11 $C_8H_{20}Pb$
8.12 CH

Quiz

1. 2243.1 g CaO
2. 393.8 g H_2O
3. C_3H_6O
4. AlF_3
5. $C_6H_{12}O_6$
6. 969.6 g N_2
7. 3598 g S
8. Molecular, empirical
9. C: 72.35%; H: 13.88%; O: 13.77%
10. No, she is not correct. The limiting reactant is the reactant that has the smallest mole-to-coefficient ratio.
11. 2357. g $SiCl_4$
12. 833.5 g SiO_2
13. C_3H_3O
14. 6
15. Cl_2
16. 5.64 moles of HCl; 205.64 g of HCl.

Chapter 9
The Transfer of Electrons Between Atoms in a Chemical Reaction

Practice Problems

9.1

Molecule	Dot structure
CH_3F	H H:C:H :F:

Valence electrons on free atom	Electrons owned by each atom
H: 1 electron	H: 0 electrons
C: 4 electrons	C: 6 electrons
F: 7 electrons	F: 8 electrons

Oxidation state	Sum
H: $1 - 0 = +1$	H: $+1 \times 3 = +3$
C: $4 - 6 = -2$	C: $-2 \times 1 = -2$
F: $7 - 8 = -1$	F: $-1 \times 1 = -1$
	$= 0$

9.2

Molecule	Dot structure
H_2O	H:Ö:H

Valence electrons on free atom	Electrons owned by each atom
H: 1 electron	H: 0 electrons
O: 6 electrons	O: 8 electrons

Oxidation state	Sum
H: $1 - 0 = +1$	H: $+1 \times 2 = +2$
O: $6 - 8 = -2$	O: $-2 \times 1 = -2$
	$= 0$

9.3

Molecule	Dot structure
ClO_4^-	:Ö: :Ö:C:Ö: :Ö:

Valence electrons on free atom	Electrons owned by each atom
Cl: 7 electrons	Cl: 0 electrons
O: 6 electrons	O: 8 electrons

Oxidation state	Sum
Cl: $7 - 0 = +7$	Cl: $+7 \times 1 = +7$
O: $6 - 8 = -2$	O: $\underline{-2 \times 4 = -8}$
	$= -1$

9.4

Molecule	Dot structure
CO	:C:::O:

Valence electrons on free atom	Electrons owned by each atom
C: 4 electrons	C: 2 electrons
O: 6 electrons	O: 8 electrons

Oxidation state	Sum
C: $4 - 2 = +2$	C: $+2 \times 1 = +2$
O: $6 - 8 = -2$	O: $\underline{-2 \times 1 = -2}$
	$= 0$

9.5 H: +1 (rule 3)
F: −1 (halide rule)
C: −2 (rule 7)

Sum of oxidation states:

H: $+1 \times 3 = +3$
F: $-1 \times 1 = -1$
C: $\underline{-2 \times 1 = -2}$
$= 0$

9.6 H: +1 (rule 3)
S: −2 (rule 7)

Sum of oxidation states:

H: $+1 \times 2 = +2$
S: $\underline{-2 \times 1 = -2}$
$= 0$

9.7 O: −2 (rule 2)
Cl: +7 (rule 7)

Sum of oxidation states:

O: $-2 \times 4 = -8$
Cl: $\underline{+7 \times 1 = +7}$
$= -1$

9.8 O: −2 (rule 2)
C: +2 (rule 7)

Sum of oxidation states:

O: $-2 \times 1 = -2$
C: $\underline{+2 \times 1 = +2}$
$= 0$

9.9 O : −2 (rule 2)
C: +3 (rule 7)

Sum of oxidation states:

O: $-2 \times 4 = -8$
C: $\underline{+3 \times 2 = +6}$
$= -2$

9.10 $CaCl_2 + Na_2O \longrightarrow NaCl + CaO$ is not a redox reaction:

Total charge:	$+2\ -2 = 0$	$+2\ -2 = 0$	
Oxidation no.:	$+2\ -1$	$+1\ -2$	
	$CaCl_2$ +	Na_2O	$\longrightarrow$
Total charge:	$+1\ -1 = 0$	$+2\ -2 = 0$	
Oxidation no.:	$+1\ -1$	$+2\ -2$	
	NaCl +	CaO	

9.11 $2\ Al_2O_3 \longrightarrow 4\ Al + 3\ O_2$ is a redox reaction:

Total charge:	$+6\ -6 = 0$	0	0
Oxidation no.:	$+3\ -2$	0	0
	$2\ Al_2O_3 \longrightarrow$	$4\ Al$ +	$3\ O_2$

9.12 $Co + Cu^{2+} \longrightarrow Co^{2+} + Cu$

9.13 $Zn + Sn^{2+} \longrightarrow Zn^{2+} + Sn$

Quiz

1. (a)

Molecule	Dot structure
HI	H:Ï: (with two dots below I)

Valence electrons on free atom	Electrons owned by each atom
H: 1 electron	H: 0 electrons
I: 7 electrons	I: 8 electrons

Oxidation state	Sum
H: $1 - 0 = +1$	H: $+1 \times 1 = +1$
I: $7 - 8 = -1$	I: $\underline{-1 \times 1 = -1}$
	$= 0$

(b)

Molecule	Dot structure
CH_2Cl_2	H :C̤̈l:C̤̈:C̤̈l: H

Valence electrons on free atom	Electrons owned by each atom
H: 1 electron	H: 0 electrons
C: 4 electrons	C: 4 electrons
Cl: 7 electrons	Cl: 8 electrons

Oxidation state	Sum
H: $1 - 0 = +1$	H: $+1 \times 2 = +2$
C: $4 - 4 = 0$	C: $0 \times 1 = 0$
Cl: $7 - 8 = -1$	Cl: $-1 \times 2 = -2$
	$= 0$

(c)

Molecule	Dot structure
SO_3	:Ö:S:Ö: :Ö:

Valence electrons on free atom	Electrons owned by each atom
S: 6 electrons	S: 0 electrons
O: 6 electrons	O: 8 electrons

Oxidation state	Sum
S: $6 - 0 = +6$	S: $+6 \times 1 = +6$
O: $6 - 8 = -2$	O: $-2 \times 3 = -6$
	$= 0$

(d)

Molecule	Dot structure
C_2H_6	H H H:C̤̈:C̤̈:H H H

Valence electrons on free atom	Electrons owned by each atom
H: 1 electron	H: 0 electrons
C: 4 electrons	C: 7 electrons

Oxidation state	Sum
H: $1 - 0 = +1$	H: $+1 \times 6 = +6$
C: $4 - 7 = -3$	C: $-3 \times 2 = -6$
	$= 0$

2. (a) H: +1 (rule 3)
   O: −2 (rule 2)
   N: +5 (rule 7)

   Sum of oxidation states:
   H: $+1 \times 1 = +1$
   O: $-2 \times 3 = -6$
   N: $+5 \times 1 = +5$
   $= 0$

   (b) O: −2 (rule 2)
   S: +2 (rule 7)

   Sum of oxidation states:
   O: $-2 \times 3 = -6$
   S: $+2 \times 2 = +4$
   $= -2$

   (c) O: −2 (rule 2)
   Ca: +2 (rule 7)
   S: +6 (rule 7)

   Sum of oxidation states:
   O: $-2 \times 4 = -8$
   Ca: $+2 \times 1 = +2$
   S: $+6 \times 1 = +6$
   $= 0$

3. (a) $B_2O_3 + 3\,Mg \longrightarrow 2\,B + 3\,MgO$ is a redox reaction:

	B_2O_3	+	$3\,Mg$	$\longrightarrow$
Total charge:	$+6\ -6 = 0$		0	
Oxidation no.:	$+3\ -2$		0	

	$2\,B$	+	$3\,MgO$
Total charge:	0		$+2\ -2 = 0$
Oxidation no.:	0		$+2\ -2$

(b) $HF + KOH \longrightarrow KF + H_2O$ is not a redox reaction:

	HF	+	KOH	$\longrightarrow$
Total charge:	$+1\ -1 = 0$		$+1\ -2 + 1 = 0$	
Oxidation no.:	$+1\ -1$		$+1\ -2 + 1$	

	KF	+	H_2O
Total charge:	$+1\ -1 = 0$		$+2\ -2 = 0$
Oxidation no.:	$+1\ -1$		$+1\ -2$

(c) $Zn + 2\,HCl \longrightarrow ZnCl_2 + H_2$ is a redox reaction:

	Zn	+	2 HCl	$\longrightarrow$
Total charge:	0		$+1 - 1 = 0$	
Oxidation no.:	0		$+1 - 1$	

	$ZnCl_2$	+	H_2
Total charge:	$+2\ -2 = 0$		0
Oxidation no.:	$+2\ -1$		0

4. (a) Zn is more active. The reaction is spontaneous as written.
   (b) Li is more active. The reaction is spontaneous as written.
   (c) Fe is more active. The reaction is spontaneous as written.
   (d) Fe is more active. The reaction is spontaneous in the opposite direction.
5. $Mn + Ni^{2+} \longrightarrow Mn^{2+} + Ni$
6. more
7. (a) +4 (b) −3 (c) −2 (d) −1
8. Total charge: 0 +2+6−8 = 0 +2+6−8 = 0 0
   Oxidation no.: 0 +1+6−2 +2+6−2 0
   $$Mg + H_2SO_4 \longrightarrow MgSO_4 + H_2$$
   H: +1 ⟶ 0 changes
   O: −2 ⟶ −2 no change
   Mg: 0 ⟶ +2 changes
   S: +6 ⟶ +6 no change
9. The ring must be made of gold because mercury is more active than gold. If the ring were silver, it would react with the mercury, losing electrons to the mercury ions and forming silver ions. Because gold is less active than mercury, then no reaction will occur.
10. No, Cu is less active than Zn.
11. *Bookkeeping method:*

Molecule	**Dot structure**
PO_4^{3-}	$\left[\begin{array}{c} :\ddot{O}: \\ :\ddot{O}:\ddot{P}:\ddot{O}: \\ :\ddot{O}: \end{array}\right]^{3-}$

Valence electrons on free atom	**Electrons owned by each atom**
P: 5 electrons	P: 0 electrons
O: 6 electrons	O: 8 electrons

Oxidation state	**Sum**
P: 5 − 0 = +5	P: +5 × 1 = +5
O: 6 − 8 = −2	O: −2 × 4 = −8
	= −3

Shortcut method:
O: −2 (rule 2)
P: +5 (rule 7)

Sum of oxidation states:
O: −2 × 4 = −8
P: +5 × 1 = +5
= −3

Cumulative Quiz for Chapters 7, 8, and 9

1. $C_6H_{12} + 9\ O_2 \longrightarrow 6\ CO_2 + 6\ H_2O$
2. 3
3. (a) $BaSO_4$ and (c) AgCl
4. $PbSO_4$
5. Molecular equation:
   $2\ HBr + Ca(OH)_2 \longrightarrow CaBr_2 + 2\ H_2O$
   Ionic equation:
   $2\ H^+ + 2\ Br^- + Ca^{2+} + 2\ OH^- \longrightarrow Ca^{2+} + 2\ Br^- + 2\ H_2O$
   Net ionic equation:
   $2\ H^+ + 2\ OH^- \longrightarrow 2\ H_2O$, which reduces to $H^+ + OH^- \longrightarrow H_2O$
6. Molecular equation:
   $Na_2CO_3 + Ba(NO_3)_2 \longrightarrow BaCO_3(s) + 2\ NaNO_3$
   Ionic equation:
   $2\ Na^+ + CO_3^{2-} + Ba^{2+} + 2\ NO_3^- \longrightarrow BaCO_3(s) + 2\ Na^+ + 2\ NO_3^-$
   Net ionic equation:
   $CO_3^{2-} + Ba^{2+} \longrightarrow BaCO_3(s)$
7. 305.8 g of Mg_3N_2
8. 303.8 g of NH_3
9. C_2H_5O
10. $HgCl_2$
11. C_3H_3O
12. K: 40.0%; Mn: 28.0%; O_4: 32.0%
13. 556.3 g of Ca
14. Br_2 is the limiting reactant.
15. 1.887 mol HBr; 152 g of HBr

16. (a)

Molecule	Dot structure
HBr	H:B̤̈r:

Valence electrons on free atom	Electrons owned by each atom
H: 1 electron Br: 7 electrons	H: 0 electrons Br: 8 electrons

Oxidation state	Sum
H: $1 - 0 = +1$ Br: $7 - 8 = -1$	H: $+1 \times 1 = +1$ Br: $\underline{-1 \times 1 = -1}$ $= \;\; 0$

(b)

Molecule	Dot structure
CH_3Cl	H H:C̈:H :C̤̈l:

Valence electrons on free atom	Electrons owned by each atom
H: 1 electron C: 4 electrons Cl: 7 electrons	H: 0 electrons C: 6 electrons Cl: 8 electrons

Oxidation state	Sum
H: $1 - 0 = +1$ C: $4 - 6 = -2$ Cl: $7 - 8 = -1$	H: $+1 \times 3 = +3$ C: $-2 \times 1 = -2$ Cl: $\underline{-1 \times 1 = -1}$ $= \;\; 0$

(c)

Molecule	Dot Structure
SO_2	:Ö::S̈::Ö:

Valence electrons on free atom	Electrons owned by each atom
S: 6 electrons O: 6 electrons	S: 2 electrons O: 8 electrons

Oxidation state	Sum
S: $6 - 2 = +4$ O: $6 - 8 = -2$	S: $+4 \times 1 = +4$ O: $\underline{-2 \times 2 = -4}$ $= \;\; 0$

(d)

Molecule	Dot Structure
C_3H_8	H H H H:C̤̈:C̤̈:C̤̈:H H H H

Valence electrons on free atom	Electrons owned by each atom
H: 1 electron C_{end}: 4 electrons C_{mid}: 4 electrons	H: 0 electrons C_{end}: 7 electrons C_{mid}: 6 electrons

Oxidation state	Sum
H: $1 - 0 = +1$ C_{end}: $4 - 7 = -3$ C_{mid}: $4 - 6 = -2$	H: $+1 \times 8 = +8$ C_{end}: $-3 \times 2 = -6$ C_{mid}: $\underline{-2 \times 1 = -2}$ $= \;\; 0$

17. (a) H: +1 (rule 3)
O: −2 (rule 2)
Cl: +7 (rule 7)
Sum of oxidation states:
H: $+1 \times 1 = +1$
O: $-2 \times 4 = -8$
Cl: $\underline{+7 \times 1 = +7}$
$= \;\; 0$

(b) O: −2 (rule 2)
Cr: +6 (rule 7)
Sum of oxidation states:
O: $-2 \times 7 = -14$
Cr: $\underline{+6 \times 2 = +12}$
$= \;\; -2$

(c) Mg: +2 (rule 5)
O: −2 (rule 2)
P: +5 (rule 7)
Sum of oxidation states:
Mg: $+2 \times 3 = +6$
O: $-2 \times 8 = -16$
P: $\underline{+5 \times 2 = +10}$
$= \;\; 0$

18. (a) $HBr + NaOH \longrightarrow NaBr + H_2O$ is not a redox reaction:

	HBr	+	NaOH	$\longrightarrow$
Total charge:	$+1-1=0$		$+1-2+1=0$	
Oxidation no.:	$+1-1$		$+1-2+1=0$	

	NaBr	+	H_2O
Total charge:	$+1-1=0$		$+2-2=0$
Oxidation no.:	$+1-1$		$+1-1$

(b) $Zn + 2\,HNO_3 \longrightarrow ZnNO_3 + H_2$ is a redox reaction:

	Zn	+	$2\,HNO_3$	$\longrightarrow$
Total charge:	0		$+2+10-12=0$	
Oxidation no.:	0		$+1\ +5-2$	

	$ZnNO_3$	+	H_2
Total charge:	$+3+3-6=0$		0
Oxidation no.:	$+3+3-2$		0

19. (a) Ca is more active; the reaction will be spontaneous in the reverse direction.
(b) K is more active; the reaction will be spontaneous in the reverse direction.

20. $Zn + Sn^{2+} \longrightarrow Zn^{2+} + Sn$

21. Zn: 0 to +2
H: +1 to 0
Cl: −1 to −1 No change

22. No, Pb is less active than Ni.

23. *Bookkeeping method:*

Molecule	Dot structure
NF_3	:F:N:F: with :F: below N (each F with lone pairs)

Valence electrons on free atom	Electrons owned by each atom
N: 5 electrons	N: 2 electrons
F: 7 electrons	F: 8 electrons

Oxidation state	Sum
N: $5-2=+3$	N: $+3\times1=+3$
F: $7-8=-1$	F: $-1\times3=-3$
	$=0$

Shortcut method:
F: −1 (halide rule)
N: +3 (rule 7)

Sum of oxidation states:
F: $-1\times3=-3$
N: $+3\times1=+3$
$=0$

24. Potassium metal is very active (second in the EMF series); thus it would be more prone to readily react with ions present in the Earth's crust. Platinum is less reactive and would not react as quickly.

25. $Co + Sn^{2+} \longrightarrow Co^{2+} + Sn$

Chapter 10 Intermolecular Forces and the Phases of Matter

Practice Problems

10.1 Dipole–dipole forces and London forces
10.2 Hydrogen-bonding, dipole–dipole forces, and London forces
10.3 CH_3OH has the higher boiling point because in addition to dipolar forces and London forces, it contains hydrogen-bonding.
10.4 $CH_3CH_2CH_2CH_3 < CH_3CH_2OCH_3 < CH_3CH_2CH_2NH_2$

Quiz

1. Dipole–dipole forces and London forces
2. CH_3OCH_3 has the boiling point of 30°C because it contains dipole–dipole forces and London forces. CH_3CH_2OH has the boiling point of 78°C because in addition to dipolar forces and London forces, it contains hydrogen-bonding from the O–H bond.
3. (a) CH_4 (b) CH_3CH_2Cl
4. (a) London forces
   (b) Dipole–dipole forces and London forces
   (c) Hydrogen-bonding, dipolar forces, and London forces
5. Higher
6. (a) $CH_3CH_2CH_2CH_3$ because it's larger and therefore has greater London forces.
   (b) CH_3CH_2OH because it has greater London forces.
7. CH_3NH_2
8. (b) CH_2Cl_2 and (c) CH_3OH
9. $CH_4 < CH_3CH_3 < CH_3F < CH_3CH_2Cl < CH_3OH$

10. CH_3OH. Although both involve hydrogen-bonding, oxygen is more electronegative than nitrogen. Therefore the O–H hydrogen bond is stronger than the N–H hydrogen bond.

Chapter 11
What If There Were No Intermolecular Forces? The Ideal Gas

Practice Problems

11.1 4.12 atm

11.2 1047 K

11.3 0.387 mol

11.4 131 g/mol

11.5 34.0 g/mol

11.6 1.73 atm

11.7 0.8 L or 800 mL

11.8 78.5 L

11.9 1.0 mole of P_4

11.10 4.67×10^4 grams of CO_2

11.11 12.6 moles of O_2

Quiz

1. 20.6 L
2. 23 mol
3. 4170 g $CaCO_3$
4. (a) 3.40 L; they have an inversely proportional relationship.
   (b) 13.6 L; they have a directly proportional relationship.
5. 10.4 L
6. 570°C
7. 4.86 L
8. 7.88 atm
9. (a) $2\ H_2O_2 \longrightarrow 2\ H_2O + O_2$
   (b) 1.21 moles of H_2O_2
10. 5.00 L of He contains 0.41 moles of gas (a); 2.50 g of He contains 0.63 moles of gas (b); thus 2.50 g of He contains more moles than 5.00 L of He.
11. 1.91×10^3 KCl
12. 30.1 g/mol

Chapter 12
Solutions

Practice Problems

12.1 Place 293 g of NaCl in a 1.00-L volumetric flask, and add enough water to bring the level of the solution to 1.00 L.

12.2 Place 21.4 g of $C_{12}H_{22}O_{11}$ in a 250-mL volumetric flask, and add enough water to bring the level of the solution to 250 mL.

12.3 Place 556 mL of stock solution in a 1.00-L volumetric flask, and bring the solution level up to the desired volume.

12.4 Place 167 mL of stock solution in a 250-mL volumetric flask, and bring the solution level up to the desired volume.

12.5 12.5% CH_3OH

12.6 11.8% $C_{12}H_{22}O_{11}$

12.7 0.50 g of SiH_4

12.8 To prepare 45g of PbC_2O_4:
Combine 608 mL of 0.250 M $Pb(NO_3)_2$ and 1.22 L of 0.125 M $K_2C_2O_4$.

12.9 0.0647 M HCl

12.10 0.0984 M H_2SO_4

12.11 176 g/mol

Quiz

1. 1.5 g of $BaCl_2$
2. Place 8.4 mL of $C_2H_2O_6$ in a 100-mL volumetric flask and bring the solution level up to the desired volume.
3. 4.7% ethanol
4. 81 mL of HCl
5. 0.0268 M H_2SO_4
6. 24.2% KNO_3
7. 20.1 g/mol
8. 40% KBr
9. 6.8% C_3H_6O
10. Place 44.81 g of Na_2SO_4 in a 500-mL volumetric flask, and add enough water to bring the level of the solution to 500.0 mL.

Cumulative Quiz for Chapters 10, 11, and 12

1. Hydrogen-bonding
2. Vaporization
3. Lower
4. (a) $CH_3OCH_2CH_3$
5. (a) Dipolar interactions, London forces, and hydrogen-bonding
   (b) Dipolar interactions and London forces
   (c) London forces
6. BF_3 is a nonpolar molecule because the individual polar bonds cancel each other due to the molecular shape.
7. 82.0 L
8. 436.8 g CaO; 187,000 mL
9. 1.14 L
10. 253 L
11. 603 K
12. 9.8 L
13. 250 g He
14. Water
15. (d) SO_2
16. 20%
17. 189 mL
18. 124 g/mol
19. 0.015 mol HNO_3
20. CH_3CH_2OH has dipole–dipole interactions, London forces, and hydrogen-bonding, whereas CH_3OCH_3 lacks hydrogen-bonding.

Chapter 13
When Reactants Turn into Products

Practice Problems

13.1 The reaction is exothermic.

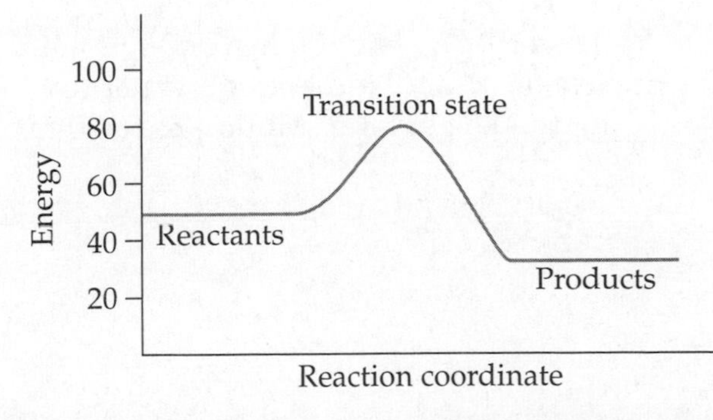

13.2 The reaction is endothermic.

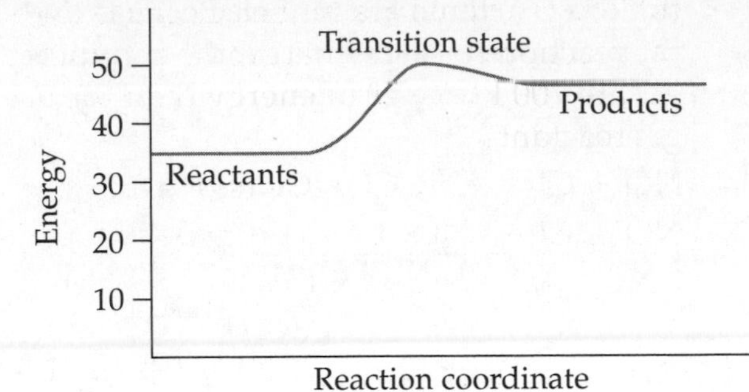

13.3 Rate = $k[R]^x[S]^y$

(a) Use experiments 1 and 2 to find x:
When [R] triples, the rate triples:
$3^1 = 3, x = 1$

(b) Use experiments 1 and 3 to find y:
When [S] doubles, the rate doubles:
$2^1 = 2, x = 1$

Thus the rate law is:
Rate = $k[R]^1[S]^1$
so the overall order is $x + y = 2$.

13.4 $Br_2 \longrightarrow 2\ Br$ (slow)
$2\ Br + CHBr_3 \longrightarrow HBr + CBr_4$
$Br_2 + CHBr_3 \longrightarrow HBr + CBr_4$
Note: Step 2 is trimolecular; improbable but possible.

Quiz

1. Reactants; products
   Products; reactants
2. (a) Rate = $k[CO]^1[NO]^1$ (b) 2
   (c) The rate will be doubled.
3. The reaction is endothermic.

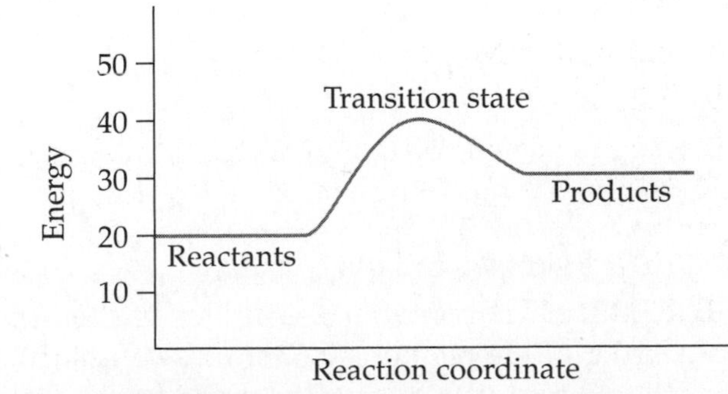

4. Rate = $k[NO]^x[O_2]^y$
5. (a) The rate will increase by a factor of 4.
   (b) The rate will be tripled.
   (c) The rate will increase by a factor of 9.
   (d) The rate will increase by a factor of 12.

6. (a) Exothermic
   (b) The reactants are higher. Because the reaction releases energy, the products are 700 kJ lower in energy than the reactants.
7. $NO + Cl_2 \longrightarrow NOCl + Cl$ (slow)
   $NO + Cl \longrightarrow NOCl$ (fast)
   $2\,NO + Cl_2 \longrightarrow 2\,NOCl$
8. 3
9. Rate $= k[H_2][ICl]$
10. Rate $= k[A]^2[B]^1$

Chapter 14
Chemical Equilibrium

Practice Problems

14.1 0.207
14.2 9.5×10^3
14.3 1.7×10^{-6}
14.4 7.9×10^{-9}
14.5 8.8×10^{-7} M

Quiz

1. 7.4×10^{-8}
2. Products, reactants
3. 2.83×10^{-11}
4. $K_{eq} = \dfrac{[SO_3]^2}{[SO_2]^2[O_2]}$
5. 0.314 M
6. 4.0×10^{-8}
7. 4.0×10^7
8. 8.0×10^{-2} M
9. 1.22×10^{-8}
10. AgCl is more soluble in water than AgBr because it has a higher concentration of Ag^+ ions. $[Ag^+]$ equals the moles of salt (AgCl or AgBr) that are dissolved per liter of solution. Because the $[Ag^+]$ is higher in the AgCl solution, AgCl is more soluble than AgBr.
11. The reaction $E \rightleftarrows F$ with a $K_{eq} = 1 \times 10^{-18}$ will contain more reactants in its equilibrium state because the $K_{eq} < 10^{-3}$, which means that the equilibrium lies far to the left. Therefore the reaction barely occurs, so mostly reactants are present.
12. $K_{eq} = \dfrac{[NH_3][H_2S]}{[NH_4HS]}$

Chapter 15
Electrolytes, Acids, and Bases

Practice Problems

15.1 (a) Strong acid; strong electrolyte
(b) Weak acid; weak electrolyte
(c) Weak acid; weak electrolyte
15.2 (a) Strong base; strong electrolyte
(b) Weak base; weak electrolyte
(c) Strong acid; strong electrolyte
15.3 K_2SO_4
15.4 HBr: acid
Br^-: base
NH_3: base
NH_4^+: acid
15.5 CH_3NH_2: base
$CH_3NH_3^+$: acid
H_2O: acid
OH^-: base
15.6 7.7×10^{-13}
15.7 1.7×10^{-7}
15.8 $H_2BO_3^- + H_3O^+ \longrightarrow H_3BO_3 + H_2O$
15.9 $HCN + OH^- \longrightarrow CN^- + H_2O$

Quiz

1. (a) Electrolyte (strong)
   (b) Nonelectrolyte
   (c) Electrolyte (weak)
2. (a) Na_3PO_4
   (b) $Mg(NO_3)_2$
3. (a) $H_2BO_3^-$
   (b) $C_2H_3O_2^-$
   (c) NH_3
4. (a) Strong acid
   (b) Weak acid
   (c) Strong acid
   (d) Strong acid
5. NaF

6. 1.0×10^{-7}
7. (a) Weak base
   (b) Weak base
   (c) Strong base
   (d) Strong base
8. (a) NH_4^+
   (b) HNO_3
   (c) $H_2PO_4^-$
9. 1.12×10^{-10} M
10. (a) and (d)
    (e) and (g)
    (b) and (f)
    (c) and (h)
11. (a) $HS^- + H_3O^+ \longrightarrow H_2S + H_2O$
    (b) $H_2S + OH^- \longrightarrow HS^- + H_2O$
12. 1.0×10^{-9}

Cumulative Quiz for Chapters 13, 14, and 15

1. The reaction is exothermic because heat is released into the environment (the container), which causes the container to be hot to the touch.
2. (a) Rate = $k[NO_2]^2[Br_2]$
   (b) 3
   (c) The rate will double.
   (d) The rate will increase by a factor of 9.
3.

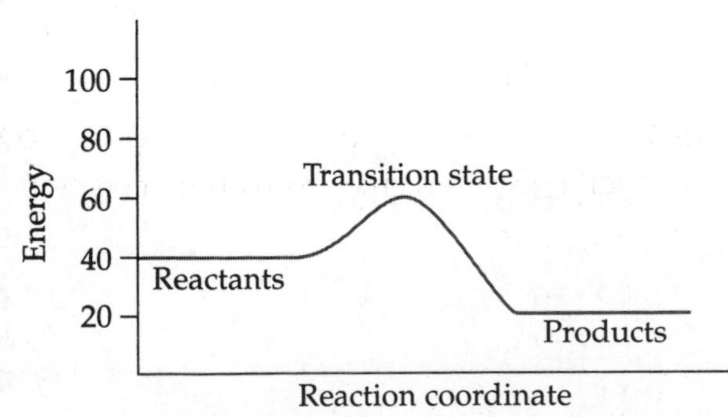

The reaction is exothermic because the energy of the products is less than the energy of the reactants.

4. Rate = $k[CH_3Br]^x[OH^-]^y$
5. (a) Endothermic
   (b) The products are higher. Because the reaction consumes energy, the products are 250 kJ higher.
6. Rate = $k[A_2][BC]$
7. Right
8. Reactants, products
9. False. The value of K_{eq} changes only if the temperature changes.
10. 1.6×10^{-10}
11. (a) $K_{eq} = \frac{[C_2H_6]^2[O_2]}{[C_2H_4]^2[H_2O]^2}$
    (b) $K_{eq} = \frac{[N_2O][NO_2]}{[NO]^3}$
12. 1.4×10^{-2}
13. $K_{eq} = \frac{[CO_2]^3}{[CO]^3}$
14. 2.53 M
15. (a) $K_{sp} = [Ca^{2+}][CO_3^{2-}]$
    (b) $K_{sp} = [Mg^{2+}][OH^-]^2$
16. 1.44×10^{-4}
17. No. If the mechanism were one step, the rate law would include $[Y_2]$.
18. (a) Strong base (b) Strong acid
    (c) Salt (d) Strong acid
19. HNO_3: acid; NO_3^- : base
    NH_3: base; NH_4^+ : acid
20. 1.32×10^{-10}

Chapter 16 Nuclear Chemistry

Practice Problems

16.1 0.5966 g/mol

16.2 The mass defect for $^{12}_{6}C$ is 0.0993 g/mol. The mass defect for $^{10}_{5}B$ is 0.069 85 g/mol. Therefore $^{12}_{6}C$ has the greater mass defect.

16.3 7.58×10^8 kJ/mol/nucleon

16.4 1.03×10^2 kJ/mol/nucleon

16.5 12.5 g will remain after 45.8 years.

16.6 It will take approximately 162 seconds for the isotope to decay to 4 g.

16.7 1.05×10^8 years
16.8 2.47×10^4 years

Quiz

1. 1.89×10^4 years
2. 6.76×10^9 years
3. 0.0633 g/mol
4. 1.42×10^8 kJ/mol/nucleon
5. 8.30×10^8 kJ/mol/nucleon
6. 15.63 g will remain after 175 years.
7. It will take approximately 30 hours for the isotope to decay to 3 g.
8. The mass defect per nucleon for $^{40}_{18}Ar$ is 0.009 31 g/mol/nucleon. The mass defect per nucleon for $^{27}_{13}Al$ is 0.008 99 g/mol/nucleon. Therefore $^{40}_{18}Ar$ has the greater mass defect per nucleon.
9. Mass defect
10. 1/256 of a radioactive isotope with a 25-year half-life will remain after 200 years.

Chapter 17
The Chemistry of Carbon

Practice Problems

17.1 Saturated: $C_nH_{2n+2} = C_8H_{18}$
17.2 Unsaturated: $C_nH_{2n+2} = C_8H_{14}$
17.3 4-methyl-2-hexyne
17.4 4-methyl-2-hexene
17.5 Ketone
17.6 Carboxylic acid
17.7 Alkyl halide

Quiz

1. 3-methyl-pentane
2. Alkane, alkene, alkyne
3. (a) Alcohol
   (b) Ether
   (c) Alkyl halide
4. (a) Saturated: $C_nH_{2n+2} = C_5H_{12}$
   (b) Unsaturated: $C_nH_{2n+2} = C_7H_{16}$
5. (a) R—OH
   (b) R—O—R
   (c) 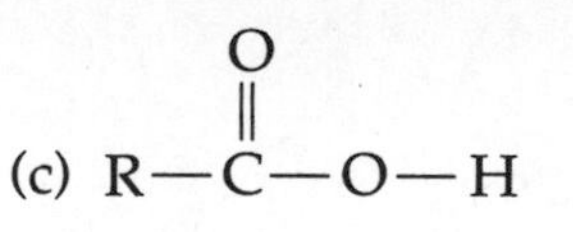
6. (a) Propane (b) Heptane
7. 4-methyl-1-hexene
8. Carboxylic acids, ketones, aldehydes
9. 26
10. The correct name should be 3-methyl-1-butene.

$CH_3CH_2CH{=}CH_3$ (with CH_3 branch)

Chapter 18
Synthetic and Biological Polymers

Practice Problems

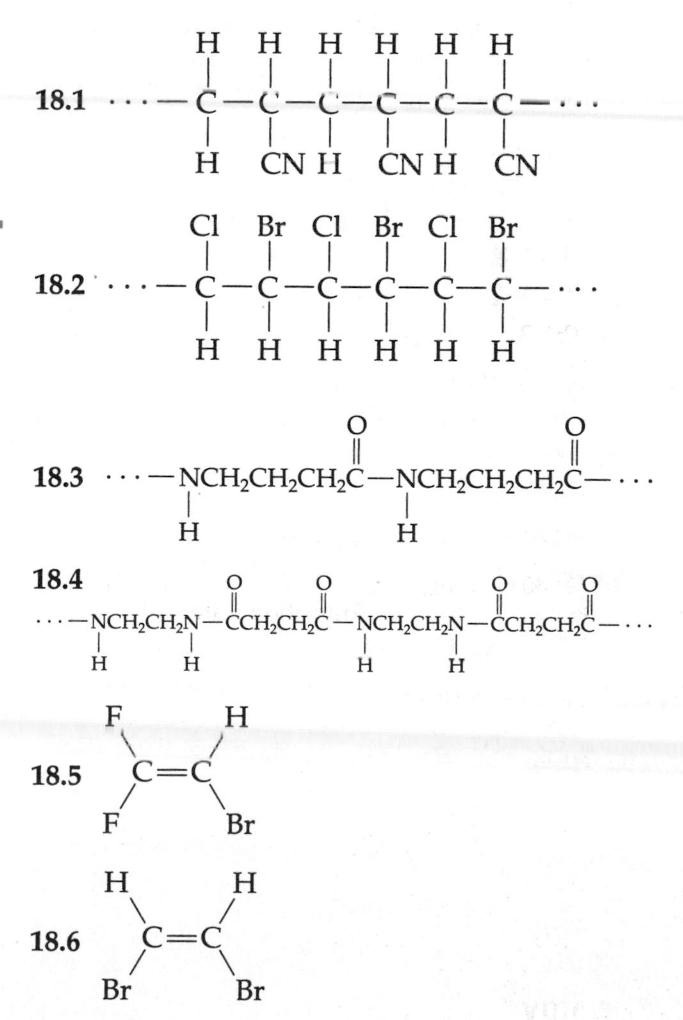

18.7 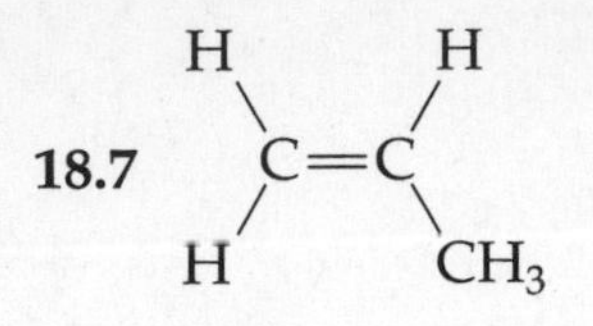

Quiz

1. and 2.

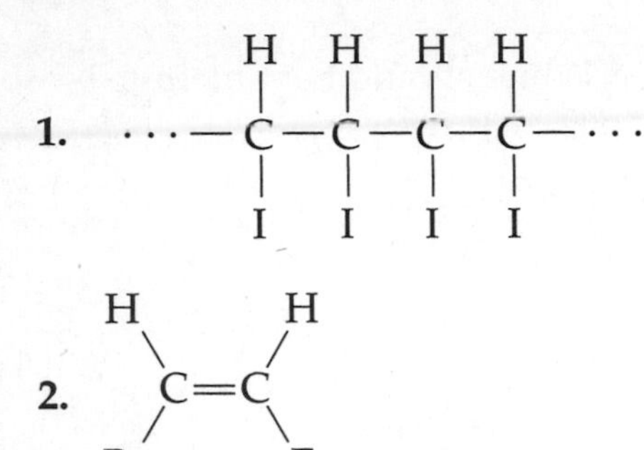

3. Amide

4. 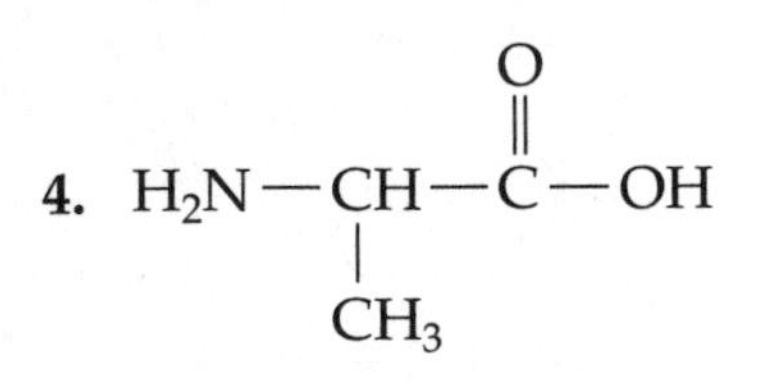

5. Water (H_2O)

6. 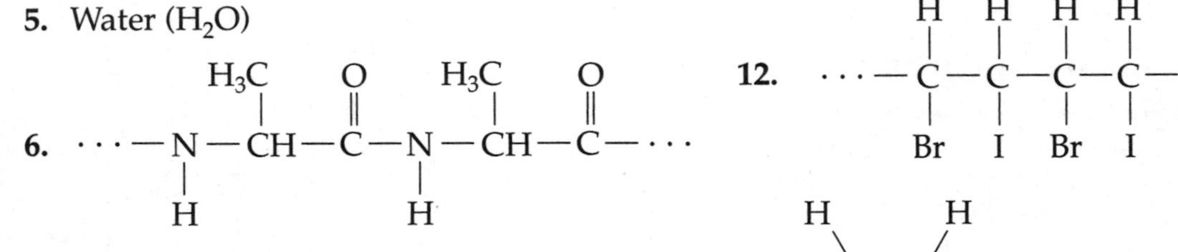

7. 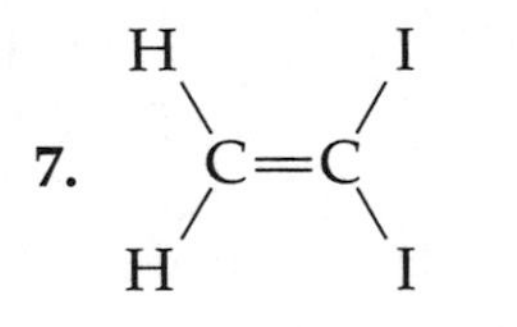

8. Double bond, single bond

9. 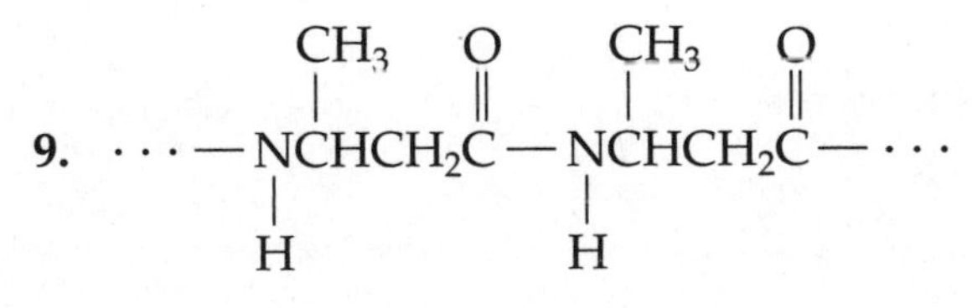

10. HCl

Cumulative Quiz for Chapters 16, 17, and 18

1. 1.33×10^4 years
2. 0.1052 g/mol
3. 7.28×10^9 kJ/mol/nucleon
4. 6.25 g
5. 5 years
6. 3-methyl-heptane
7. (a) Alcohol
   (b) Carboxylic acid
   (c) Alkyl halide
8. (a) Unsaturated: $C_nH_{2n+2} = C_4H_{10}$
   (b) Saturated: $C_nH_{2n+2} = C_5H_{12}$
9. (a) Hexane (b) Octane
10. 3-methyl-1-butene
11. $CH_3CHCH{=}CH_2$ (with CH_3 on the second carbon)

    3-methyl-1-butene

12. and 13.

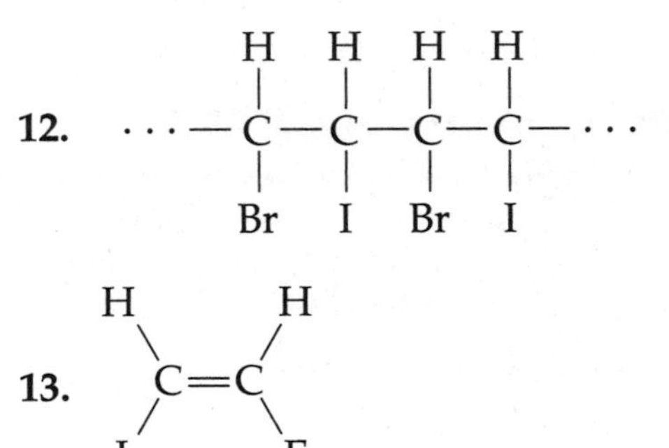

14. double, single

15. 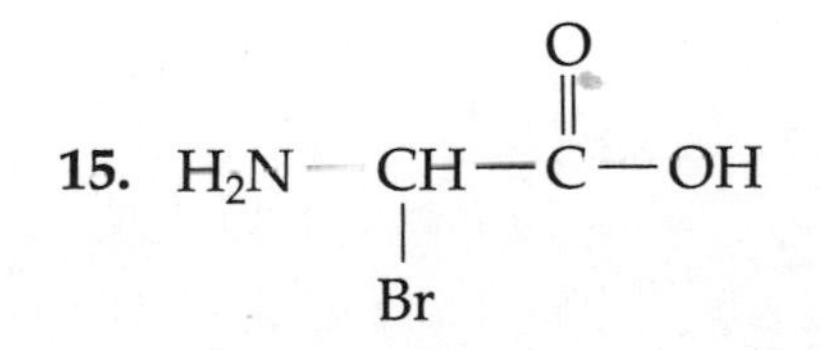

16. 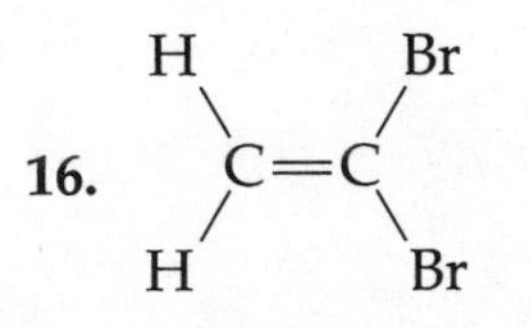

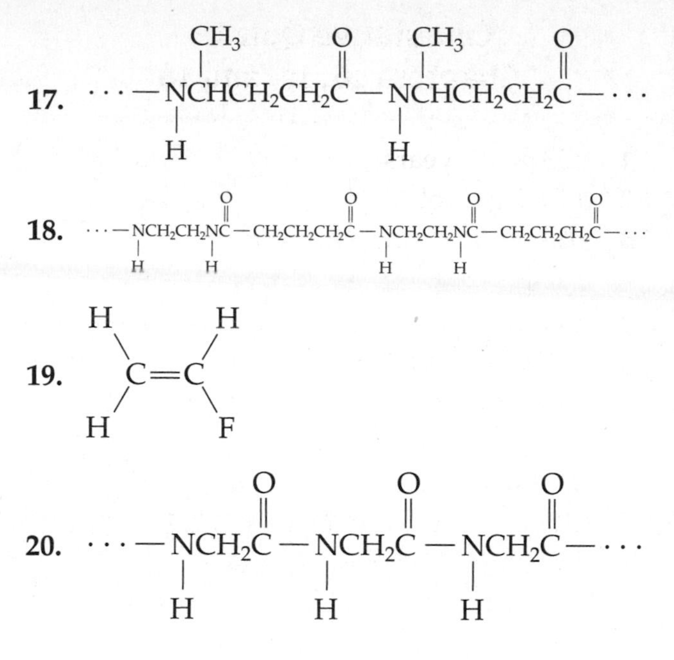

Appendix: Math Review

A.1 10

A.2 90

A.3 −4

A.4 0.60; 60%

A.5 86%

A.6 240%

A.7 135 grams of iron

A.8 2174 pounds of air

A.9 $x = 8$

A.10 $n = \dfrac{PV}{RT}$

A.11 10^{-8}

A.12 5.6873×10^{4}

A.13 8.94×10^{-5}

A.14 (a) 10^{7} (b) 10^{-1}

A.15 4.25×10^{9}

A.16 2.1×10^{6}

A.17 1.6×10^{-5}

A.18 2.0×10^{5}

A.19 4

A.20 1×10^{-9}

A.21 (a) 12:60 or 1:5
(b) 60:12 or 5:1

A.22 230 grams of sodium

A.23 125 m^3

A.24 10.81

A.25 50°C

A.26 6 atm

A.27 34,564 cm

A.28 180°

Additional resources available for instructors:

Benjamin/Cummings Digital Library for Introductory Chemistry, 2nd edition (0-321-04634-X)

An easily searchable resource containing all the visuals from the text. Images can be downloaded in a variety of formats and inserted into PowerPoint slides or other presentation tools. Instructors can also download images from this CD-ROM to their own class-secure Web site to assist them in teaching online.

Testing Program: Printed Test Bank (0-321-40557-9); **TestGenEQ Win/Mac** (0-321-05328-1)

By David A. Kort of Mississippi State University

A complete testing package with more than 700 questions that correspond to major topics in the text. Available in print or electronic version.

Benjamin/Cummings Custom Laboratory Program

Create a custom laboratory manual by selecting labs and reorganizing experiments from our existing library, or by adding your own.

Instructor's Teaching Guide for Introductory Chemistry (0-321-05332-X)

By Saundra Yancy McGuire of Louisiana State University

Provides instructors with insights and tools for presenting chemical principles to students with little or no previous exposure to chemistry. Features an overview and outline of each chapter of the text. Presents information concerning common student misconceptions and topics requiring special attention, and offers complete descriptions of appropriate lecture demonstrations on the material covered in each chapter.

Instructor's Manual for the Introductory Chemistry Laboratory Manual (0-321-05330-3)

By Doris Kimbrough of University of Colorado at Denver and Wendy Gloffke of Cedar Crest College

For each lab, the manual provides teaching instructions, variations on experiments, thought-provoking discussion questions, and postlab suggestions.

Complete Solutions Manual for Introductory Chemistry (0-321-05331-1)

By Saundra Yancy McGuire of Louisiana State University

Contains comprehensive solutions to all problems in the text.

Color Acetates (0-321-05329-X)

Includes 150 full-color acetates.

Additional resources available for students:

Introductory Chemistry Laboratory Manual (0-321-04639-0)

By Doris Kimbrough of University of Colorado at Denver and Wendy Gloffke of Cedar Crest College

Innovative labs help students to develop basic lab skills, understand lab safety, learn to collect, organize and analyze data, and exercise critical thinking. For use with the text or as a stand-alone resource.

The Chemistry of Life CD-ROM for Introductory Chemistry (0-8053-3121-2)

By Robert M. Thornton, University of California, Davis

Through high-quality animations and interactive simulations, this CD-ROM helps students master crucial chemistry concepts. Simulated lessons, complete with interactive quizzes, cover: atomic structure, reactions and equilibrium, properties of water, acids and bases, and the structure and function of macromolecules. It also includes an illustrated glossary and topics correlated with the text.

The Chemistry Place, Special Edition for Introductory Chemistry (http://www.chemplace.com/intro/russo)

Directed by Joe March of the University of Alabama, Birmingham

This unique study tool offers interactive tutorials, practice quizzes, and additional study aids, all written specifically to accompany the text.